Pass Your Driving Test
The 1990s Guide for Learner Drivers

A comprehensive and structured learning programme for learner drivers, broken down into stages that match the first stages of The Associated Examining Board's Learner Driver Progress Checks. Every practical skill is clearly demonstrated with the aid of illustrations, and the theoretical knowledge you must acquire is introduced in natural stages, with self-check sheets and questions at each learning stage.

Learner drivers and their teachers can use this manual to ensure that every lesson, every practice, consolidates each new skill, so that you will be a skilled driver for Life.

Peter Russell, as well as being a writer and broadcaster on motoring matters, is General Secretary and Training Officer of the Driving Instructors Association (DIA), and is an experienced teacher and driving instructor himself.

He is actively involved with The Associated Examining Board on the Diploma in Driving Instruction and their Learner Driver Training and Testing programmes, and has been responsible for the preparation of their syllabus. He is also a member of the Parliamentary Advisory Council for Transport Safety (PACTS) and Technical Editor of *Driving* magazine.

Peter Russell

Pass Your Driving Test

The 1990s Guide for Learner Drivers

Pan Books

London, Sydney and Auckland

First published 1992 by
Pan Books Ltd, Cavaye Place, London SW10 9PG

9 8 7 6 5 4 3 2 1

© Peter Russell 1992

ISBN 0 330 31830 6

Printed and bound in Great Britain by
BPCC Hazells Ltd
Member of BPCC Ltd

Permission to reproduce the Statement of Failure DL24 on pages 148 and
149 of this book has been granted by the Driving Standards Agency,
with the approval of the Controller of Her Majesty's Stationery Office,
and is acknowledged with thanks by the publishers

Contents

Pupil's personal progress check

When subject introduced: ☑   Name: _____

When subject is confirmed: ☒

1 Licence checked: ☐

   Signed: ☐   Category: ☐   Expires: /   /

2 Eyesight checked: ☐   Spectacles/contact lens: ☐

3 Highway Code received: ☐   Tested: ☐   OK: ☐

4 Cockpit drill: ☐   Starting procedure: ☐

5 Controls: Accelerator ☐   F/brake ☐   Clutch ☐

            St Wheel ☐   H/brake ☐   Gears ☐

            Mirror and door mirrors adjusted ☐

6 Minor: Ign/start ☐   Indicators ☐   Wipers ☐

          Horn ☐   Lights ☐   Washers ☐

          Speedo ☐   Fuel/oil ☐   Rear/wipe ☐

          Heater ☐   De-mister ☐   H.R.W. ☐

7 Move away level: ☐   Move away gradients: ☐

8 Stop normally: ☐   Emergency stop: ☐

9 Start; steer; stop safely (without instruction): ☐

10 Turn left into: ☐   Out from: ☐

11 Turn right into: ☐   Out from: ☐

12 Xroads with priority: ☐   Without priority: ☐

13 Roundabouts: Left ☐   Across ☐   Right ☐

14 MSM procedure: ☐   P.S.L. ☐   L.A.D. ☐

15 Mirrors used correctly: ☐

16 Signals given correctly: ☐

17 Positioning correct R & L turns: ☐

18 Observation correct at all junctions: ☐

19 Correct action at Tfc Lights; Signs; Controllers: ☐

20 Proper use of speed: ☐

21 Proper progress: ☐

22 Overtake: ☐   Meet: ☐   Cross path of: ☐

23 Normal position: ☐   Without shaving: ☐

24 Pedestrian crossings: ☐   Uncontrolled: ☐   Controlled: ☐

25 Safe stopping places: ☐

26 Awareness and anticipation: ☐

Each check point to be ticked only when driver can cope unaided.

Date Test booked for: /   /   Pass final mock test: /   /

Instructor's Comments:

Acknowledgements: The Author and Publishers wish to thank
The Associated Examining Board for kind permission to reproduce samples
from their Learner Driver Progress Check and for use of their logo;
the Driving Instructors Association for permission to use their logo; and
Her Majesty's Stationery Office for permission to reproduce the syllabus
from *Your Driving Test*.

Introduction

How to use this book

A book cannot teach you to drive, but it can help you learn many of the peripheral things which concern all drivers. With the changes to the driving test in Britain for the first time in over fifty-five years, brought about by the Harmonization Directives of the European Community, it is now more than ever important that all Learner drivers understand much more about the theoretical aspects of being a driver on today's roads.

The first four parts of this book provide a structured learning programme which matches those of The Associated Examining Board Learner Driver Progress Checks and gives instructors and pupils the total theoretical background that is required of any competent driver. Each part is separate in itself, and yet fits into the whole pattern of learning by objectives. These objectives are based on the Recommended Syllabus for Learning to Drive (see pages 15–26) and at the end of each part there is a check-list of those skills you should have mastered by this stage. In addition to the specific learning objectives, subsidiary information and guidance on such topics as legal requirements and the function of various car parts are included where appropriate. Part Five covers the Test itself, and Part Six provides a wealth of information and advice that will be invaluable to the first-car owner.

Where the objectives are wholly practical, such as in-car and driving skills, it is advisable that professional training from a first-class Approved Driving Instructor is used. There can be no substitute for professional training, and although some pupils may find it suitable and necessary to have private practice with parents and friends, it is stressed that this should be confined to practice only. If you find that practice with colleagues or private supervisors clashes with that given by your professional instructor, you would be well advised to give up the private practice.

Driving techniques, and indeed driving test requirements, have changed considerably over the past few years. It is not only that your professional instructor will be more likely to ensure that you pass the driving test first time, but also because the skills, knowledge and attitudes learned from a professional instructor will allow you to learn to drive as a life skill; a skill which will enable you to play your part in road safety, to the benefit of yourself, your future passengers, and all other road users.

Any prospective learner driver would do well to read this book before booking driving lessons. Note particularly the sections on 'Choosing a Professional Instructor', 'Legal Requirements' and 'Know Your Car'. A quick exploratory scan through Parts One to Five will clarify exactly what you should be expected to learn. This way you can double-check that you are getting value for money. Once you have booked a course of lessons, you can then refer to this book, ticking off those skills you have mastered. There is much to be said for doing this *with* your instructor (most teachers welcome a positive, interested response from their pupils, and are open to suggestions). However, if you are too shy to do so (or feel that you are going to offend your instructor) you can use the checklist at the end of each lesson to trace your own progress.

The licence to drive which you will receive when you pass your test of driving competence will authorize you to drive anywhere in the European Community, and in most other parts of the world, for the rest of your life. Learn how to do this well. This book is designed to give you all the background resource material you will need to do that.

At the end of the first five parts of this book, you will find a confirmation list of the objectives you should have achieved. When you (with your instructor, if you are using this book together) have ticked them all off, you can congratulate yourself that you are now one of Europe's safest drivers.

Author's note: For simplicity, I have referred to your instructor throughout as 'he'. There are of course just as many highly competent female instructors and examiners.

The new Euro driving test

Motoring is no longer an insular sport. A licence to drive in any one country is seen to be a licence to drive anywhere in the world. Since Britain joined the Common Market in 1971 the legal need for harmonization of driving licences, driver testing and driver training has been obvious. And as a direct consequence of the Treaty of Rome signed in 1980 there has been a European commitment and directive aimed to harmonize all three by the end of this century. Anyone who has been issued with a driving licence since June 1990 will have noted that even the categories under which licences are issued have been changed.

The harmonization of driver testing follows soon; this will mean that driving in Britain will certainly require a much greater knowledge of the theoretical aspects of driving as well as being able to pass a practical test. The object of this is to bring British drivers into line with the rest of Europe from a theory point of view, and to bring Europe into line with Britain from a practical point of view.

There is no doubt that the British driving test, which has remained virtually unchanged since its inception in 1935, has stood the test of time. This is not to say that it is just as easy to pass now as it was fifty or twenty years ago. The reason it has become more difficult with the passing of time is quite simply due to the volume of traffic which now exists. Today's learner drivers have to cope with much heavier traffic, but under road conditions which are made as easy as possible by the elimination of the majority of accident and danger spots. Priorities are much better understood now than ever before. Road layouts have improved enormously, and because there is no longer the potential for confusion, it is much simpler to recognize who has 'right of way' at any particular junction or place on the road. The success of this has been that although traffic densities have doubled and trebled over the past few years, the numbers of road users killed and injured per road user mile have diminished enormously.

Britain is proud of the fact that whenever road safety figures are discussed as a whole we are much better at keeping our motorists alive on the roads than our neighbours and partners in the European Community. What is not so acceptable is that, because of our attitude towards road user behaviour and the lack of training and testing on these attitudes, we manage to kill and maim many more cyclists, children and pedestrians than in other countries.

One way in which this may be changed for the better is to encourage all drivers, but especially new drivers, to understand the theoretical requirements of driving and behaviour on the road towards all our fellow users. It is not enough to prove we won't injure them in the half an hour or so that the practical test lasts; we should also be required to pass a stiff theoretical examination to show that we have acquired the correct knowledge and have developed proper attitudes towards road-user behaviour too.

It is to this purpose that the following book is devoted. The subject material is that of the recommended syllabus for all learner drivers published by the Department of Transport, and the validation of the book is covered by The Associated Examining Board's Learner Driver Progress Checks (see pages 8–9) which are now available for all professional driving instructors to use with their pupils.

The whole range of driver education in the broader sense is divided into five separate headings:

Part 1: Before you go onto the road;

Part 2: Early on in your course of tuition;

Part 3: In the middle of your course of tuition;

Part 4: Late on in your course, leading up to your driving test (the author has decided to split this section into two parts – 'Leading up to your driving test' and 'Taking your Test');

Part 5: After you have passed your driving test.

By using this textbook in association with these test papers you and your professional instructor can ensure that you will be able not only to pass any driving test, whether verbal, written or practical, but will also become a safe, competent and thinking driver.

The Department of Transport's recommended syllabus for learning to drive

Driving is a life skill, and it will take you many years to gain full mastery of the skills set out here.

This syllabus lists the skills in which you must achieve basic competence in order to pass the Driving Test.

You must also have

- a thorough knowledge of the Highway Code and the motoring laws;
- a thorough understanding of your responsibilities as a driver.

This means that you must have real concern, not just for your own safety, but for the safety of all road users, including pedestrians.

If you learn with an Approved Driving Instructor (ADI), make sure he or she covers this syllabus fully.

1 Legal requirements

To learn to drive you must

(i)   be at least 17 years old. If you receive a mobility allowance for a disability you may start driving a car at 16.

(ii) be able to read in good daylight, with glasses or contact lenses if you wear them, a motor vehicle number-plate

- 20.5 m away;
- with letters 79.4 mm (3.1 inches) high.

(iii) be medically fit to drive

(iv) hold a provisional licence or comply with the conditions for holding a provisional licence (see leaflet D100*)

(v)   ensure that the vehicle being driven

- is legally roadworthy
- has a current test certificate, if over the prescribed age
- is properly licensed with correct disc displayed

(vi)   make sure that the vehicle being driven is properly insured for its use

(vii) display L plates which are visible from the front and the back of the vehicle

(viii) be accompanied by a supervisor who
- has held a full UK licence for at least 3 years for the kind of vehicle being used;
- is at least 21 years old

(ix)   wear a seat belt, unless granted an exemption, and see that all the seat belts in the vehicle and their anchorages and fittings are free from obvious defects

(x)   ensure that children under 14 are suitable restrained by an approved child restraint or an adult seat belt

(xi)   be aware of the legal requirement to notify medical conditions which could affect safe driving. If a vehicle has been adapted for disability, ensure that the adaptations are suitable to control the vehicle safely.

(xii) know the rules on the issue, presentation or display of

- driving licences; • insurance certificates; • road excise licences

2 Car controls, equipment and components

You must

(i)   understand the function of the

- accelerator • clutch • gears • footbrake • handbrake
- steering

and be able to use these competently

(ii) know the function of other controls and switches in the car that have a bearing on road safety and use them competently

(iii) understand the meaning of the gauges and other displays on the instrument panel

(iv) know the legal requirements for the vehicle

(v) be able to carry out routine safety checks that do not require tools, and identify defects especially with

- steering ● brakes ● tyres ● seat belts ● lights ● reflectors
- direction indicators ● windscreen wipers and washers ● horn
- rear view mirrors ● speedometer ● exhaust system

(vi) know the safety factors relating to vehicle loading

3 Road user behaviour

You must

(i) know the most common causes of accidents

(ii) know which road users are most at risk and how to reduce that risk

(iii) know the rules, risks and effects of drinking and driving

(iv) know the effect of fatigue, illness and drugs on driving performance

(v) be aware of age dependent problems among other road users especially among children, teenagers and the elderly

(vi) be alert and be able to anticipate the likely actions of other road users and be able to suggest appropriate precautions

(vii) be aware that courtesy and consideration towards road users are essential for safe driving

4 Vehicle characteristics

You must

(i) know the most important principles concerning braking distances and road holding under various road and weather conditions

(ii) know the handling characteristics of other vehicles with regard to stability, speed, braking and manoeuvrability

(iii) know that some vehicles are less easily seen than others

(iv) be able to assess the risks caused by the characteristics of other vehicles and suggest precautions that can be taken, for example

- large commercial vehicles pulling to the right before turning left
- blind spots for drivers of some commercial vehicles
- bicycles and motorcycles being buffeted by strong wind

5 Road and weather conditions

You must

(i) know the particular hazards of driving

- in both daylight and dark
- on different types of road, for example
  – on single carriageway, including country lanes
  – on three-lane roads
  – on dual-carriageways and motorways

(ii) gain experience in driving in urban and higher speed roads (but not motorways) in both daylight and darkness

(iii) know which road surfaces provide the better or poorer grip when braking

(iv) know the hazards caused by bad weather, for example
- rain • fog • snow • icy roads
- strong cross winds

(v) be able to assess the risks caused by road and traffic conditions, be aware of how the conditions may cause others to drive unsafely, and be able to take appropriate precautions

6 Traffic signs, rules and regulations

You must

(i) have a sound knowledge of the meaning of traffic signs and road markings

(ii) have a sound grasp of the traffic signs, for example
- speed limits
- parking restrictions
- zebra and pelican crossings

7 Car control and road procedure

You must have the knowledge and skill to carry out the following tasks
- in both daylight and darkness
- safely and competently
- making proper use of mirrors, observation, and signals

(i) take necessary precautions before getting into or out of the vehicle

(ii) before starting the engine
   - carry out the 'cockpit drill' including fastening the seat belts
   - take proper precautions

(iii) start the engine and move off
   - straight ahead and at an angle
   - on the level and on uphill and downhill gradients

(iv) select the correct road position for normal driving

(v) take proper observation in all traffic conditions

(vi) drive at a speed suitable for road and traffic conditions

(vii) react promptly to all risks

(viii) change traffic lanes

(ix) pass stationary vehicles

(x) meet, overtake and cross the path of other vehicles

(xi) turn right and left, and at junctions, including crossroads and roundabouts

(xii) drive ahead at crossroads and roundabouts

(xiii) keep a safe separation gap when following other vehicles

(xiv) act correctly at pedestrian crossings

(xv) show proper regard for the safety of other road users, with particular care towards the most vulnerable

(xvi) drive on both urban and rural roads and, where possible, dual carriageways – keeping up with the traffic flow where it is safe and proper to do so

(xvii) comply with traffic regulations and traffic signs and signals given by the police, traffic wardens and other road users

(xviii) stop the vehicle safely, normally and in an emergency, without locking the wheels

(xix) turn the vehicle in the road to face the opposite way using the forward and reverse gears

(xx) reverse the vehicle into a side turning keeping reasonably close to the kerb

(xxi) park parallel to the kerb while driving in reverse gear

(xxii) park the vehicle in a multistorey car park, or other parking bay, on the level, uphill and downhill, both in forward and reverse direction

(xxiii) cross all types of railway level crossings

8 Additional knowledge

You must know

(i) the importance of correct tyre pressures

(ii) the action to avoid and correct skids

(iii) how to drive through floods and flooded areas

(iv) what to do if involved in an accident or breakdown, including the special arrangements for accident and breakdown on motorways

(v) basic first aid for use on the road as set out in the Highway Code

(vi) the action to take to deter car thieves

9 Motorway driving

You must gain a sound knowledge of the special rules, regulations and driving techniques for motorway driving before taking your driving test.

After passing your test you should take motorway lessons with an ADI before driving unsupervised on motorways.

10 Points for riders of motorcycles and mopeds

You must master everything in sections 1 to 9, except the items which clearly do not refer to you.

In addition a learner rider must

(i)   know the safety factors relating to safety helmets and how to adjust the helmet correctly

(ii)   know the safety factors in wearing suitable clothing and in using goggles or a visor

(iii)   know the importance of rear observation
   ● by use of mirrors
   ● by looking over the shoulder
   ● including the life-saver look

(iv)   know how to lean while turning

(v)   be able to carry out additional safety checks for two-wheel vehicles, for example
   ● chain tension and condition
   ● condition of control cables
   ● steering-head play
   ● suspension
   ● wheels, and tightness of all nuts and bolts

(vi)   be able to use the front and rear brakes correctly

(vii)   be able to keep the machine balanced at all speeds

(viii) be able to make a U-turn safely

(ix)   be able to wheel the machine, without the aid of the engine, by walking alongside it

(x)   be able to park and remove the machine from its stand

Department of Transport recommended syllabus for learner drivers

Practical requirements, broken down into achievable **objectives** in ten stages of training.

Note: Parts 1–4 in this manual take you precisely through these ten recommended stages.

Date	Subject	Ability				Comments
	Able to enter the vehicle safely and carry out all safety checks					
	Able to move away safely (level)					
	Able to focus ahead, steering to follow a *Safety Line*					
	Able to brake smoothly, and able to bring the vehicle to rest at a designated spot on the road					
	End of Stage 1					
	Able to identify and explain the use and operation of the main controls of the vehicle					
	Able to enter, start the engine, and move off from rest without any instruction					
	Able to observe road hazards whilst on the move					
	Able to select all gears as required (including 'block' changes) without looking					
	Able to bring the car to rest, safely every time, as required. And also quickly, but with safety, as in an emergency					
	Able to move away on any hill (up and down)					

Date	Subject	Ability				Comments
	Able to move away at an angle up to half a car's length from any obstruction					
	Able to indicate (and signal by arm) while on the move, safely					
	End of Stage 2					
	Able to recognize, identify and cope with each of the following: bend, corner, turning, junction, crossroad, and roundabout (mini, normal and large)					
	Able to negotiate each of the above, safely and correctly, on approach, while on them, and leave them correctly and safely: to the left, right and ahead					
	End of Stage 3					
	Able to recognize, identify and cope with all types of road and traffic signs, while driving in light traffic conditions					
	Able to identify motorway signs; and explain their meanings					
	Able to show a high standard of knowledge of the Highway Code, and other motoring matters					
	End of Stage 4					

Date	Subject	Ability				Comments
	Able to reverse in a straight line					
	Able to reverse round a corner to the left, on level ground					
	Able to turn the vehicle in the road, using forward and reverse gears; safely and correctly, with due regard to all other road users					
	And accurately with regard to various widths of road					
	End of Stage 5					
	Able to reverse round a corner to the right, on level ground					
	Able to reverse to the left and right, while on all types of gradient					
	Able to turn the vehicle round in a confined space (such as a car park), correctly and safely, under full control, correctly positioned and with maximum observation and care for all other road users					
	End of Stage 6					
	Able to park the vehicle, using reverse gear, in any suitable confined space					
	Between other vehicles					
	At a parking meter bay					
	To the right and left					
	End of Stage 7					

Date	Subject	Ability				Comments
	Able to understand and answer questions on night driving, adverse weather conditions, and cope with them as they arise					
	Able to cope with all kinds of urban and rural road traffic conditions					
	End of Stage 8					
	Able to understand and answer questions on dual carriageways, and motorway driving procedures					
	Able to reassure the Instructor that motorway driving will be coped with safely and correctly after the test is passed					
	End of Stage 9					
	Able to cope with a simulated Driving Test route					**With no serious or dangerous errors**
	Able to show effective use of mirrors					
	Able to manoeuvre as required for the test					
	Able to demonstrate making good progress					
	Able to meet all other traffic safely and correctly					
	Able to maintain a correct *safety line*					
	Able to act correctly at pedestrian crossings (of all kinds)					
	Able to show awareness and anticipation of the actions of all other road users					
	Able to answer questions on any motoring matters, satisfactorily for the standard required of a new driver					
	Able to identify any road or traffic sign and demonstrate the correct reaction					

Date	Subject	Ability	Comments
	Able to maintain full control of the vehicle at all times; and to take charge of the traffic situation as required. Making and taking all the required decisions in good time, and correctly End of Stage 10		
	Post-Driving-Test Instruction Able to drive safely and with confidence on urban and rural motorways, entering, driving along and leaving them correctly and safely Able to overtake other traffic, making correct use of lanes, and making suitable headway as necessary for road and traffic conditions at the time Able to drive to a standard acceptable to the Institute of Advanced Motorists, or the RoSPA Advanced Drivers Association Able to cope with a simulated Advanced Driving Test, lasting at least 90 minutes		

Choosing a professional instructor

Driving is a life skill and as such is not something to be mastered completely as part of a standard 'L driver training course'. Nevertheless what can be learned in such a course is the foundation for that life skill. Perhaps the most important thing for every learner driver to do initially is the selection of their instructor who will train them to pass their driving test, which is a step towards gaining the necessary experience to become a safe and competent driver. There is no way that anyone can learn the physical and practical skills of driving a motor car whilst reading a book. On the other hand there are so many incidental aspects of learning, that to leave the understanding and acquisition of these areas of knowledge to be picked up while having practical training means that either they will never be learned, or the time taken to learn them will prove quite expensive.

This book will ensure that learner drivers have learned all that is needed to make them safe, competent drivers who recognize the need to treat driving as if their very lives depended upon it.

It goes without saying that as well as understanding all the requirements of the driving test all drivers should have a good working knowledge of the *Highway Code*. This booklet is not something that should be studied for a few frantic hours the night before you take your test. It is a guide for all people who, in whatever capacity, need to use the roads at any time, and needs to be read and studied to supplement your training *throughout* your lessons.

As well as the *Highway Code*, there is another booklet called *Your Driving Test* (HMSO). This, too, should be studied, because although the need to 'learn to drive as a life skill' is the aim, one of the important objectives to be achieved on the way is to pass the driving test.

As a brand-new learner driver you will need to select your driving instructor with great care. You may like to ask your family and friends who taught them to drive, though if you study the way some of your friends drive you might be tempted to look for an instructor who didn't teach them. The choice of an instructor can be made a little bit easier by looking in the Yellow Pages, or Thomson Blue Pages, to see what sort of advertisements instructors have put there to tempt you to learn to drive with them. The first thing you want to know is what sort of qualifications your instructor has. The fact that he is an Approved Driving Instructor should be taken for granted. It may be that you could be offered tuition by a 'Trainee Instructor'; and it could well be that some trainee instructors do offer excellent training. Indeed they have to practise on someone.

However, learning to drive is so important that if you do have a trainee, make sure that *your* lessons are the ones which are supervised. (Trainees need to be supervised for a fifth of their lessons. Insist that your lessons are the ones which *are* supervised.) Other qualifications may or may not have significance, but are worth looking at. Membership of a Trade Association, for instance, has no real teaching merit, although it will offer a Code of Professional Practice to help you if any disputes should arise. Membership of the

Institute of Advanced Motorists can also be taken for granted. Every professional instructor would be able to pass such a test quite easily, though DIAmond Advanced Instructors are granted automatic exemption from this advanced driving test, as are Department of Transport Driving Examiners. Nevertheless, ask if your instructor has passed the advanced driving test, and which ones?

Other qualifications which your instructor should or could have include possession of the Diploma in Driving Instruction; the RAC Registered Instructor's Certificate or a Trade Association's Teaching Certificate. The highest qualification which can be held in the driver training world is the DIAmond Advanced Instructor. Such a qualification can only be held by an instructor who has passed a very stringent (5 separate 2-hour written modules) theoretical examination for the Diploma in Driving Instruction set by the Associated Examining Board, and an equally stiff practical examination set by the Department of Transport's own Examiner Training Establishment.

Once again, ask your instructor what qualifications he or she holds, and whether they were passed by examination, or simply offered as part of their membership of an association. One question which you can ask in an effort to find out more about your instructor is the simple one 'Have you been to Cardington?' Cardington is the home of the Department of Transport's Driving Examiner Training Establishment. All DIAmond Advanced Instructors will have been there for their own examination, but all professional driving instructors are invited to visit Cardington to see how Driving Examiners are trained and what standards are looked for in the driving test.

Every driving lesson should start with a 'briefing' from your instructor. The briefing should be – as you would expect – brief; but it should detail exactly what he expects you to learn by the end of that lesson. By using a book such as this, you will find that the whole subject of learning to drive is broken down into simple, readily achievable 'objectives', each of which is a stepping-stone on the way to learning to drive. Each lesson should therefore begin with something like: 'At the end of this lesson you will be able to . . .' If necessary you could ask your instructor what the lesson will include.

If this doesn't prompt him to give you a proper briefing, then perhaps you should look for an instructor who will.

Incidentally, over the past few years the Registrar of Driving Instructors has brought in some much more stringent rules for the inclusion of instructors on the Government Register. When it was first established, the Register of Approved Driving Instructors (ADISs) was content to allow instructors on it if they didn't give wrong or illegal instruction. Nowadays, any instructor who does not ensure that actual learning takes place can be removed from the Register. So by asking your instructor 'What will I be able to do by the end of this lesson?' you are probably helping him to give you a better lesson.

Driving instructors are now graded by the Department of Transport. The majority of them are Grade 4. Anyone who is Grade 3 or below must improve or they are taken off the Register. About 25 per cent or so are Grade 5 instructors. This means they are above average and well worth looking for. A few, about 5 per cent or possibly less, are Grade 6 instructors. Well worth seeking out, but don't worry too much if you cannot find one. Although instructors are allowed to advertise what grade they are, not all of them do; but if an instructor claims a particular grade it can be checked with the Department of Transport's Registrar.

Careful selection of a good instructor, and one who uses a structured training programme similar to that contained in this book, will give you excellent value for money, and this has to be what you are looking for. If you choose a cheap driving instructor, one who does not value his own skills and knowledge very highly, then do not be surprised if the instruction you receive is also cheap and ineffective. Everyone deserves a good professional driving instructor. Failure to get one does not only lead on to failure of the driving test; it can also lead on to developing unsafe driving habits, which can prove much more expensive than professional driving lessons.

Over 90 per cent of learner drivers take some professional training with an Approved Driving Instructor these days. Unfortunately, many of them do not have quite enough. They try to skimp by having just three or four too few. By using this book you can make sure that you have absolutely the right number to get you through safely; simply check that you have achieved every one of the objectives detailed below. Learning to drive is an investment for life. The money that you spend on professional training will be much less than you will ever spend on insurance premiums or damage repairs. Spend just enough of it with the right professional instructor and you will save yourself a great deal of money, and probably quite a lot of anguish too.

Legal requirements

Before getting into the driving seat of a motor car for the very first time you need to be sure that you comply with all the various legal requirements that a learner driver is obliged to obey. You must possess a provisional driving licence.

Obtaining one is simple. Go to almost any Post Office and ask for forms D100 and D1. Complete the latter and send it off to First Application Section, Swansea, SA99 1AD, with a postal order or cheque for £21. It is better to send a postal order whenever you deal with any Government agency or department. No action is ever taken until they have safely banked your fee. A cheque takes time to cash so they tend to hang on to your application form until it is cleared. A postal order means they can respond immediately.

Your application will be for a provisional licence, and as soon as the licence is received, check that the details shown, including your date of birth, are all correct and then sign it straight away. Forgetting to sign it is not only an offence, it means that if you turn up for your driving test with the licence unsigned your test will be cancelled before you even start it. Check too that the licence categories shown are correct. All licence groups were changed to categories, and the letters adjusted to match, in June 1990. If you are learning to drive a car the category is now 'B'. The date of expiry should be that of your seventieth birthday. After that you will need to renew it on a three-yearly basis, unless a medical condition prevents you from driving.

Your eyesight needs to meet the minimum standard of being able to read a conventional number-plate at 20.5 metres in good daylight. You must also be at least 17 years old, unless you are in receipt of the Government Mobility allowance which allows you to start your driving lessons at the age of 16 instead.

Whilst you do not need a medical certificate, if you are in any way uncertain about how various medical conditions may affect your driving you should consult your doctor and your driving instructor. You may not drive unaccompanied, unless you live on one of a very few Scottish islands which allow learners to drive on their own.

The vehicle you are driving must comply in all respects with the Construction and Use Regulations concerning Motor Vehicles and, whether or not you are the owner or the keeper of the vehicle, as the driver you are officially responsible for the condition and safety of that vehicle. It must be fitted with standard red L-plates on a white 7-inch square background. The L-plates must be properly fixed and perpendicular. A roof L-plate – provided it complies with regard to shape, size and colour and is clearly visible – is perfectly acceptable. The vehicle also needs to be taxed, insured for you to drive, and also insured for the carriage of a Government Driving Examiner at the appropriate time. The tax disc must be properly displayed in the bottom or top left-hand corner of the windscreen. The certificate of insurance, and also the MoT certificate if the vehicle is more than three years old, must be available for inspection if required. If you are stopped at any time by the police, you may be required to produce all the above certificates and licence at a police station of your choice within seven days. Failure to do so can be used as evidence that you or the vehicle do not comply with the law, and an offence is committed.

The person accompanying you at any time must have held a full British driving licence for that category of vehicle for at least three years, and must be at least twenty-one years of age. You and the supervisor must also both wear seat belts if they are fitted, and as the driver you are fully responsible for any children under the age of 14: they must wear seat belts or be seated in proper child-seats whether in the front or rear passenger seats, where these are fitted.

Although your supervising driver is legally required to be with you, in law you are regarded as the driver of that vehicle and can be held responsible for any infringement of the law, by you or by that vehicle. The supervising driver could also be held liable for an offence of aiding and abetting. In cases where the supervising driver is also a professional Approved Driving Instructor, any offence proved against you would almost certainly affect his own registration as an instructor. He will, therefore, most certainly do all that he can to avoid your being in danger of infringing any motoring laws.

If the vehicle is your own, or belongs to someone other than the instructor or the school, it is essential that you as the driver have written confirmation that the vehicle is properly insured for professional driver training. Most private insurance policies are for 'Social Domestic and Pleasure' purposes. This can quite often specifically exclude the giving of driver training for professional purposes. As the driver you would be held completely responsible for any irregularities.

Needless to say, every professional Approved Driving Instructor, teaching in his own car, would be completely responsible for the insurance of his own vehicle while you are at the wheel. One of the weaknesses of present domestic insurance policies is that they allow members of the family to be covered for most driving risks, yet the carrying on of a business under a Social, Domestic and Pleasure policy is usually excluded. If you do take lessons in your own or your family car, you must check.

If you are taking training with anyone else other than an ADI you need to inform your insurance company and check what cover is given to you. Quite often they will quote 'Act Only'. This is the minimum insurance cover possible and does not cover you for any personal injury, nor for any repairs to your car, nor possibly any other vehicle you may become involved with. If a friend or colleague offers to teach you for any payment whatever – even petrol money, or occasional payments – not only is it illegal but it can negate your insurance cover.

One final point on insurance cover: insurance for driving instruction is usually about double that of normal policies. A few driving schools make a habit of covering their insurance premiums through a special fee which they charge additionally to the hire of the car for driving test purposes. This is perfectly acceptable, as the insurance has to be paid for in some form or other. However, it smells a little of sharp practice when the customer is not told this until just before the driving test when it would be too late to find another driving school. It might be a good idea to ask your instructor, right at the beginning of the lessons, exactly what payments need to be made. Similarly, some pupils assume that because they have to pay for the examiner to sit with them they don't have to pay for the use of the car. Unfortunately you do. But once more this is the sort of thing that you should settle with your instructor before the lessons commence.

Know your car

Major controls

The controls of a motor vehicle are relatively standard, regardless of the type or make. This certainly applies to the main controls which come under three basic headings:
foot controls; hand controls; eye controls.

There are three **foot pedals** (for 'automatic cars', see page 180).

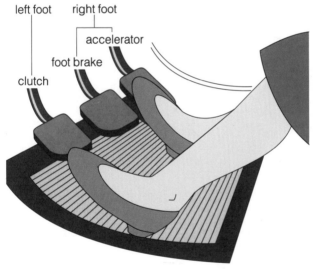

The three foot pedals

The right foot controls the accelerator and brake pedals; the left foot controls the clutch pedal.

The **accelerator pedal** controls the speed of the engine and is pressed against a light return spring. It is very sensitive and only slight pressure can give you all the control you need.

The **brake pedal** controls the speed of the car, slowing it down by applying pressure on all four wheels. Because of the way weight is transferred when braking takes place, more pressure is applied to the front wheels than the rear ones. Most cars are fitted with disc brakes on the front wheels these days. They are more effective and efficient than drum brakes, which are usually fitted to the rear wheels. Drum brakes are sometimes fitted to all four wheels on some cars. The difference between cars fitted with drum brakes all round and discs/drums is quite noticeable and it is always worth while getting used to the brakes of any new car by trying them out gently before driving off.

The **clutch pedal** is a means of joining the engine to the wheels. The clutch itself is a device fitted between the engine and the gearbox, and serves two separate functions. It allows the gears to be selected without noise. And it enables the driver to move off smoothly and to stop the car without having to stop the engine.

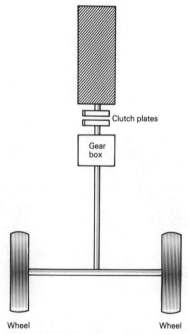

The clutch is fitted between
the engine and the gear box

There are also three **hand controls**.

The **steering wheel** is operated by both hands, and normally both hands must remain on the wheel. The purpose of the wheel is to enable the driver to follow a safety line, and also turn corners.

The **handbrake** is a parking brake, and is only used for this purpose. It should never be operated while the vehicle is moving. Its purpose is to keep the vehicle stationary and not to slow it down. Very occasionally cars are fitted with a foot-operated parking brake but even these usually have a form of hand-operated release mechanism.

The **gear lever** is used to select whichever gear is needed. It is used in conjunction with the clutch. There are usually four or five forward gears and they form the shape of an extended H pattern. There is a tendency to increase the number of gears these days, and a number of cars are now being fitted with up to seven gears in order to make them work more efficiently. The most common gear box configuration is shown on page 64.

The three areas which are covered by the **eyes** are not controls in the strict sense of the word, but because they are essential parts of the controls lesson it is useful to see how they fit into the pattern of threes.

The **windscreen** shows the road ahead and must be kept clean and clear at all times.

The **mirrors** enable the driver to be aware of some (but not all) of what is happening around him. They cover the road behind and partly at the side. All new cars have driver's door mirrors fitted as standard. Passenger door mirrors are also vital. However, a mirror is only as good as the person using it.

The **blind spots** are the areas not covered by windscreen and mirrors. A useful exercise at the beginning of learning to drive is for the instructor to walk round the car while the pupil sits in the driving seat, looking ahead and through the mirrors. The blind spots are then made very apparent. It is in these areas that potential danger exists, especially when you are about to change speed or direction. In some circumstances it is essential that blind spots are physically and visually checked before any action is taken. If it is not possible to check, always assume there is someone there.

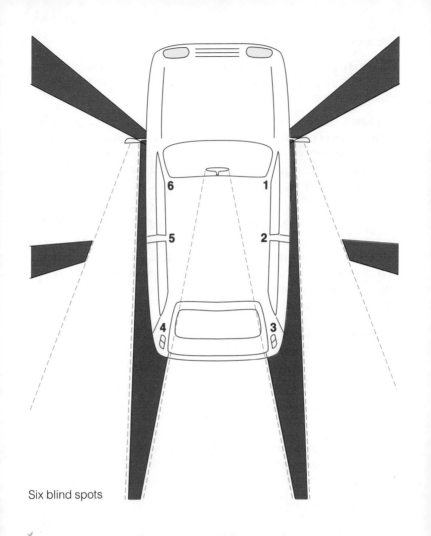

Six blind spots

Starting in reverse or first gear can move the car
enough to injure or kill a child

The function, the use and the operation of the controls

Each pupil must know three things about each of the controls: what they do; how they are used; and the words which will be used by the instructor to direct the pupil to operate them.

As a general guide the vocabulary is straightforward, but of course each instructor is free to use words which he feels are most suited to his own style. Whichever words are used the instructor and the pupil need to establish this common vocabulary at the outset. For instance, the accelerator pedal is often called a 'gas' pedal by some instructors. They like to keep the word short and simple, and gas is short for gasoline. No matter what word you both use, it should always be used. The operating words need to be kept short and simple too.

SET THE GAS; MORE GAS; LESS GAS; OFF GAS are examples of precise vocabulary that require little explanation. Words like 'a little gas', however, are too imprecise.

COVER THE BRAKE; GENTLY BRAKE; GENTLY BRAKE TO STOP are again all self-explanatory and unlikely to cause any confusion.

COVER THE CLUTCH; CLUTCH DOWN; SLOWLY CLUTCH UP also cover the clutch pedal instructions.

STEER TO THE RIGHT/LEFT; FOLLOW A SAFETY LINE; TURN THE WHEEL TO THE RIGHT/LEFT These may need a simple explanation, but once understood are unlikely to lead to any confusion.

APPLY THE BRAKE; PREPARE THE BRAKE; RELEASE THE BRAKE all clearly define what is to be done, and avoid any confusion between the footbrake and the parking brake.

HAND ON THE GEAR LEVER; PALM TOWARDS ME/YOU; SELECT FIRST/SECOND/THIRD/FOURTH/FIFTH/REVERSE GEAR These words enable the instructor to explain precisely how to position your hand on the gear lever, initially; and then the word 'select' will also become the prompt word during subsequent lessons for the pupil to expect to be asked to carry out a gear change.

The actual words used by your instructor will be his decision of course; but the above words have wide acceptance already in the profession and avoid any confusion. Certain phrases should never be used.

Control	What it does	How it does it	Words to be used
Accelerator	controls engine speed	feeds petrol to engine	Set gas
			More gas
			Less gas
			Off gas
Footbrake	controls wheel speed	applies friction to all four wheels	Cover brake
			Gently brake
			Gently brake to stop
Clutch	joins engine to wheels	separates and joins two plates connected to the engine/gearbox	Cover clutch
			Clutch down
			Slowly clutch up
Steering wheel	enables the car to change direction	turns the front wheels right and left; or to keep straight ahead	Follow safety line
			Steer (to R or L)
			Turn the wheel (to R or L)
Handbrake	holds rear wheels still	locks rear wheels with a ratchet	Apply handbrake
			Prepare handbrake
			Release handbrake

What the words mean

Press the accelerator pedal enough to enable the engine to cope. Your instructor will demonstrate precisely how much is needed.

Increase the engine speed.

Decrease the engine speed.

Take your foot off the accelerator pedal, prepare to brake.

Put your foot over the footbrake, but do not press it yet.

Commence braking progressively; the actual amount of pressure needed depends upon your speed.

Continue braking progressively until the car gently comes to rest. Only put down the clutch just prior to stopping/stalling.

Put your foot over the clutch pedal; do not press it down until necessary.

Press the clutch completely down to the floor.

Allow the clutch pedal to come up, **slowly**, until you can feel the engine note change when stationary. Still bring your clutch pedal up slowly, even when changing gear, but allow any speed difference to be taken up smoothly.

Move the steering wheel enough to follow the shape of the road, keeping both hands on the wheel.

Move the wheel with both hands on the wheel, but not more than half a turn.

Use your hands to move the steering wheel more than half a turn, by passing the wheel through your hands in turn. Always keep one hand pulling or pushing.

Pull the handbrake (or parking brake) fully on, but avoiding any ratchet noise by holding down the button.

Press the handbrake (or parking brake) ratchet button completely in so that the brake is ready to release. But still holding the brake on with your hand.

Push the handbrake (or parking brake) completely down to the floor. Check that the brake warning light also goes out.

Control	What it does	How it does it	Words to be used
Gear lever	Selects gear from choice of 4 or 5 + R	enables engine speed to be used economically	Hand on gear lever Palm to instructor Palm to pupil Select 1, 2, 3, 4, 5, R
Windscreen	enables driver to see road ahead	must be clean and clear	Look ahead
Mirrors	enables driver to see road behind	must be correctly adjusted	Check mirrors
Blind spot	limits driver's vision	need to check over shoulder	Check blind spots (physical check) Full observation

When your insert the ignition key into the lock and turn it slowly, you will feel it go through the four stages:

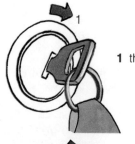

1 the steering wheel is unlocked;

2 the auxiliary circuit is switched on, which means you can have items like the radio on without the engine running;

3 the two warning lights will come on:

What the words mean

Be prepared to change gear.

See what is happening ahead.

Look in each of your mirrors for other road users.

Check those areas behind you which cannot be seen in the mirrors.

Check the road ahead, behind, blind spots and over your shoulders for any other road users who may be affected by what you are about to do.

the ignition warning light which only comes on when you are driving to warn 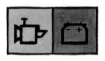 you that the electrical system is not charging properly;

the oil warning light which also only comes on when you are driving to warn you that the oil pressure has dropped or that you are very low on oil.

The reason these lights come on every time you switch on the engine is to show that they are working. If they come on while you are driving you will need to pull in and check what the fault is before continuing.

4 The final turn of the key will start the engine. Turn the key, listen to the engine as it fires and when it starts, release the key, which automatically returns to the previous position.

'Stop' for instance is never used by Driving Examiners except for the emergency stop exercise. Otherwise they will always require you to 'pull in' at a defined spot on the road. Instructors would therefore benefit from asking their pupils to 'pull in' instead of 'stop' in the same circumstances, for two reasons: it will condition the pupil to the sort of vocabulary which Driving Examiners use; and it also prevents the pupil from misinterpreting the action required.

The vocabulary used between instructor and pupil will always be determined by the instructor; nevertheless it is essential that the words chosen are both agreed and accepted to have only one precise meaning.

Minor controls

Once you have switched on the engine you can try all the other controls. Without the engine running you can't try them all. First of all, the **ignition switch** itself is usually spring loaded. You turn the key clockwise once to switch on various items like the radio and electric windows. The second position lights up the ignition warning light and various gauges such as petrol and temperature; the third actually fires the starter motor. Once the engine fires and starts to run, release the key which will return to the stage 2 position. Leave it there until you need to switch off the engine.

The most needed of the minor controls is undoubtedly the **direction indicator switch**, which is usually found alongside the steering wheel so that it can be operated without taking your hand from the wheel. On some cars it will be found on the right of the steering wheel column, on others on the left. But whichever side it is, the way you signal is always the same. Turn the indicator switch the same way that the wheel will turn after you have signalled. Try it and see. Get your instructor to show you how it works. Also see how the self-cancelling system for the indicators works. Bear in mind that there will be occasions when it won't cancel and these must be obvious to you. On other occasions it could be that it will cancel before you want it to; when that happens be prepared to switch it on again. Find out where the indicator tell-tale light is so that you can keep half an eye open for it when you are driving.

The other instruments and lights you need to watch for when you are driving are the **speedometer**, the **oil warning** and **oil pressure lights/gauges**, and the **petrol gauge/warning light**. Your instructor will take you carefully through each of the gauges and lights on his particular car; but once more remember that when you buy your own car you'll have to learn them all again, this time on your own, with the aid of your car manual.

The **windscreen wipers** and **washers controllers** are usually on the opposite side of the steering wheel to the indicators. Once again, get your instructor to demonstrate which switches they are and how they are operated. The same applies to the **lighting switches** and to the **rear fog lights**, the **heated rear window switch** and the various **heater controls**. At this stage of your learning you can afford the luxury of letting your instructor remind you when they are needed and even, at the very beginning, actually operate them for you. But by the second stage of your lessons you will need to be able to operate them yourself. And by the time you have applied for your driving test you should be pretty confident about being able to use them every time you think they are needed. One thing is certain, the Driving Examiner won't tell you to put them on or off.

The minor controls of the motor car are quite important. However, because your concentration is needed on the main controls, you can allow your instructor to determine when you are ready to take over full control of them yourself.

Finally, assuming you are learning to drive in a driving school car, there will also be a set of pedals and extra mirrors on the instructor's side. Usually these consist of footbrake and clutch only, with one or two extra mirrors so that the instructor can not only control the car but has a good idea of what is happening around and behind too. One word of caution: the use of dual controls is normally only so that the instructor may take over control of the car in an emergency. Good instructors do not use the dual pedals to help the pupil. They use them to avoid embarrassment to others. In fact the hallmark of the novice instructor is that he wants to play with the dual controls. The experienced instructor rarely uses his, because he knows exactly how much to make you work, and will not let you get out of your depth.

Part 1 Introduction to driving as a skill

At the end of this, the first stage of your driver training, you will be able to:
Enter the vehicle safely and carry out all the safety checks;

Move away safely on the level;

Focus ahead, steering to follow a safety line;

Brake smoothly and bring the vehicle to rest at a designated spot on the road.

Objectives: Department of Transport Syllabus: Stage 1

Once your practical driver training starts you will be asked to listen to what your instructor tells you to do, and put his teachings into practice. Ideally each stage of learning will be split into three separate phases: first the actual lesson to be learned will be explained and possibly demonstrated. Then you will practise it with the instructor, gradually reducing the amount of tuition and prompting that you receive, until finally you are able to demonstrate to your instructor, and all other road users, that you are now able to perform that particular operational manoeuvre on your own without any further guidance.

That is when the third phase is completed. This is learning by objectives. Your instructor, and this book, will define each of the objectives to be achieved by you, practically; and you have to learn how to cope and then demonstrate your ability to achieve the stated objective. Although it seems that there are thousands of specific practical things you are required to do before you are allowed to drive on your own, in fact they can be broken down into the following series of stated objectives. When you can achieve all of them, you will also have reached your aim which is to drive safely, and unaccompanied, while you continue to learn to drive on your own.

The aim of your professional driver training is to be able to drive on your own, safely and unaccompanied, anywhere you wish to go. Each of the following **objectives** is a stepping stone on the way to achieving that **aim**.

The first objective is simple: you must be able to:

> Enter the vehicle safely and
> carry out all the safety checks

There are two safety checks. The first, known as 'cockpit drill', takes place every time you enter the car. The second needs to be carried out before you switch on, or restart, the engine.

First safety check or 'cockpit drill'

First, you must remember exactly how you have been taught to open the car door, so that the safety of all other road users is taken into

Ideally, you should be sitting square in the seat with your bottom firmly pressed into the seat and backrest. In this illustration, the learner is sitting in the *correct* position. She can push all three pedals right down to the floor without moving her body forward and, at the same time, she can hold the steering wheel lightly but firmly.

account. Check that the road is clear of traffic, open the door wide, and get in by sliding your bottom into the seat and bringing your legs into the footwell. Try to close the door safely as soon as you can. Remember that car doors have a double catch to them, so it is possible for you to think that the door is closed when it is only on the first latch. Make sure it is properly closed by pulling it to with your *left* hand across your body. Confirm with any passengers that their doors are also secured. Then check that the handbrake or parking brake is properly applied. Your extra weight sitting in the car might be enough to move a car that hasn't got the brake on.

Adjust your seat to suit your own leg length and body height. Most seats have two adjustments: one for the fore-aft position of the seat, and one for the rake of the back rest. Some older or cheaper cars only have one adjustment. Others have an extra adjustment for the seat height as well. If you ever get into the luxury class of vehicle you can buy some that have an electronic memory for your position. But until the day you buy one, you'll have to remember to adjust it yourself instead!

Ideally your bottom needs to be well into the base of the seat, with your left foot able to press the left pedal completely down on the floor and your right foot able to reach both the other pedals. Your hands should be able to circle completely round the steering wheel without any obstruction. Once you are sure your seat is comfortable, and is firmly fixed, then put on your seat belt. This should reach across your shoulder and across your waist, fixing into the catch usually found between the front seats. Check it for comfort, and if this is the first time you've worn a seat belt or it feels different, also check how you take it off.

Finally, in your initial safety cockpit drill, make sure that you can see properly through each of the mirrors. The centre mirror should be adjusted for the maximum vision of the road behind you, without needing to move your head at all. Your door mirrors should show you the edge of the car and what is approaching you on the sides.

To help you remember what is involved in cockpit drill, your instructor will probably give you a mnemonic. Perhaps the best one is: D, treble S, M, which stands for: **doors – seat, steering, seat belt – mirrors**

There are others. Let your instructor tell you how he wants you to remember it and say the sequence out loud until it has been thoroughly learned. In the first few lessons you can let your instructor remind you, then only prompt you when required. Once you have learned it by heart there will be no need for him to tell you ever again. The objective has been achieved. This is a sequence you will use for the rest of your life; ensure that it becomes a habit.

Second safety check

The next sequence concerns starting the engine. Before you start or restart the engine of your car you must *always* check that the handbrake or parking brake is firmly applied. Then check to feel that the gear lever is in the neutral position. Only then should you turn the key in the ignition to start the engine. In the first few lessons your instructor will explain how you can be sure that the handbrake is fully applied and that the gear lever is in the neutral position. But after two or three lessons he will expect you to show him instead.

Once you have switched on the engine it is possible to see how much fuel there is. It is up to your instructor at this stage to explain to you how much fuel is recorded on the gauge. However, once you have passed your test, possibly even before you own your first car, you will have to be responsible for checking (and buying) your own petrol or diesel. Incidentally, now is a good time to ask your instructor if his car runs on petrol or diesel. You are unlikely to guess otherwise.

Handbrake	Gear lever	Starter

If your instructor teaches this way, he will not give you the key to start the engine at the beginning of each lesson until you have demonstrated both of the above sequences to him, without being asked. So if you are sitting in the car waiting to start and you still haven't been given the key, it may be that you haven't carried out both safety checks correctly!

The second objective is more practical and requires absolute safety.

Move away safely on the level

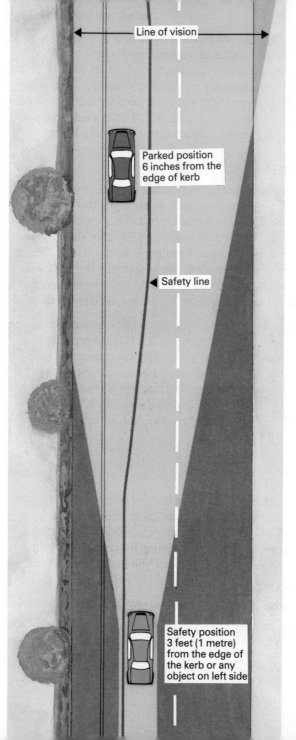

Line of vision

Parked position 6 inches from the edge of kerb

Safety line

Safety position 3 feet (1 metre) from the edge of the kerb or any object on left side

The safety position and safety line

This requires proper co-ordination of the accelerator pedal, as the clutch comes up, with releasing the parking brake at the precise moment you want. One way to demonstrate that you are moving away exactly when you want to is to be able to signal by arm at the same time, since the car must be ready to pull away, but without it moving, while you make your signal and then replace your hand on the steering wheel. Moving away on the level is not too difficult, and in your early lessons you will move off quite easily without any fear of rolling back. At a later stage you will also have to demonstrate your ability to move off on a hill.

Focus ahead, steering to follow a safety line

The **safety position** is the place where the car is, relative to the kerb and any other traffic, in the safest possible place to be, and is about a door width (or a metre) from any kerb or parked vehicle. If you are moving, the safety position becomes a **safety line** (see opposite). This is the path taken by any vehicle which is the safest possible position at all times. As you will see, sometimes it requires the driver to get back into the left after passing a series of parked cars, and on other occasions it is preferable to stay out for a longer stretch.

The correct hand position for steering

Needless to say, at no time must you be closer to oncoming vehicles on the other side of the road than to stationary (and probably empty) vehicles on the left. If for any reason you do not have enough room to be at least one metre from both stationary traffic on your left and oncoming traffic on the right, you must slow down or wait.

The safety line must always be a smooth one. Where there are obstructions in the road ahead, such as parked vehicles or narrower sections of road, try to adjust your steering gently so that the movement of the front wheels is a gradual one. The secret of safe, smooth steering is to look far enough ahead so that your brain tells your fingers what to do in plenty of time. That way your steering will always be so gentle that passengers will never be aware of any movement at all.

In order to steer correctly you will need to keep your hands in the correct position on the wheel. If you imagine the shape of the steering wheel as a clock face, in order to steer effectively your hands should be in the 10 and 2 o'clock positions. It may be necessary or more comfortable for some drivers to hold it at the 9 and 3 o'clock positions instead, which is allowable. If you find neither of these holds comfortable you are probably in the wrong seating position. Adjust your seat a little forward or back until you can hold and maintain a correct and comfortable steering wheel grip.

Steering consists of two separate types of activity:

- keeping your hands steady on the wheel while you move it gently in order to follow the line of the road;
- turning the wheel, when you need to turn a corner to join another road, or negotiate a sharp bend in the road you are on.

Ideally, your instructor will always tell you whether he requires you to *steer* – that is, to make minor adjustments to your position on the road which do not require turning the wheel – or to *turn the wheel*.

> Brake smoothly and bring the vehicle
> to rest at a designated spot on the road

Braking smoothly requires you to take your right foot away from the accelerator pedal and place it firmly on to the brake pedal, pressing the pedal sufficiently to make an appreciable effect on the speed at which the car is travelling. Initially, most learner drivers – due no doubt to nerves – tend to be too fierce on the footbrake. They soon learn, however, that at very slow speeds a gentle touch on the pedal can stop the car quite suddenly, only to find that the same touch has very little effect when travelling at more speed. The secret is to vary the pressure on the brake pedal according to the speed at which you are travelling. The faster you are going, the more pressure you will need to slow down; the faster you are going the longer it takes to stop. At 30 m.p.h. it would take the distance you can read a number plate to stop, even in an emergency.

The hallmark of the good driver is that he sees the need to brake early enough not to need to brake sharply. In fact, throughout your driving lessons and during the driving test you should only ever need to brake sharply when you practise 'emergency stops'. More about these later.

In your early lessons, you will need to practise stopping at a precise place chosen by your instructor. He will say such things as 'Bring the car to rest alongside that lamppost'. In order to impress him with your skill, you should aim to stop the car gently alongside the lamppost with his door mirror exactly beside it. If you always manage to stop in this precise manner you will not only please your instructor, you will have learned a very valuable lesson which will be of benefit for the rest of your driving life.

To recap on starting and moving away from the kerb

When the engine is not running

Check that the handbrake is applied; that the gear lever is in neutral; and then start the engine.

When the engine is running

Clutch down; hand on gear lever, palm towards instructor.

Select first gear.

Set gas (press the accelerator pedal enough to give sufficient power to the engine).

Check mirrors, over your shoulder for the blind spot and road ahead. If clear, check mirrors again.

(Signal – indicator right – only if instructor wishes.)

Slowly clutch up until engine note changes; **feet still**.

Release hand brake; both hands on the steering wheel.

Steer to the right, and then left, to follow a safety line.

Slowly clutch up completely, check mirrors again.

Cancel indicator if used.

More gas.

To pull in:

Check mirrors (including passenger door mirror if fitted).

Signal with left indicator (if instructor tells you to).

Off gas; cover clutch and cover brake.

Steer slightly left and then right to a parking position.

Gently brake, then gently brake to a stop. Clutch down.

Feet still. Apply handbrake; select neutral.

Cancel indicator if used.

Relax feet.

Ideally, your instructor will talk you through the complete exercise the first time. The second time he will use 'prompt' words to remind you of the sequence. The third time you will be able to carry out the exercise unaided. In practice, it usually takes more than three attempts to complete the sequence correctly. But by carrying out this and every other exercise in these three phases (instruction, prompted practice and validation) you will be able to carry out each exercise completely unaided quite soon.

Once you are able to complete the exercise on your own, your instructor should never need to give you instruction on it again. However, it must be accepted that a certain amount of prompting may be needed on occasions, especially where the external traffic conditions may distract you.

Achievement of objectives

As you draw to the end of each part of this training book, you will need to confirm (to yourself) that you have achieved the objectives laid down at the commencement of each part. You can either complete this check sheet as a personal commitment to yourself, or you can invite your instructor to confirm your ability to perform each of the tasks to the standard which is acceptable to him or her.

At the end of the first part of the book you will be able to:

Enter the vehicle safely, and carry out all the safety checks;

Move away safely on the level;

Focus ahead, steering to follow a safety line;

Brake smoothly and bring the vehicle to rest at a designated spot on the road.

Signed:

Pupil: _____

Instructor: _____

Date: _____

Questions

Part 1

The questions which follow are based on the multiple response principle; that is to say, a question is asked, and several answers (a), (b), (c), (d) & (e) are suggested. One or more of the answers are correct, therefore it is necessary to say (√) yes or (×) no to each of the answers. When used in conjunction with professional driver training with an Approved Driving Instructor the answers you give to the questions should be used as the basis for discussion to ensure that your knowledge of the subject matter is satisfactory.

For example:

On the driving test you may be tested on the following exercises or driving manoeuvres:

(a) Stopping as in an emergency:	(a) This is correct.
(b) Reversing or parking:	(a) This is correct.
(c) Driving on a motorway:	(a) This is wrong; you are not allowed on them.
(d) Driving in the rain:	(a) It may not rain of course, but you must expect that it might and know how to use the screen washers/wipers.
(e) Driving at speeds up to 70 m.p.h.	(a) Yes, you may well be asked to drive on a dual carriageway with a 70 m.p.h. speed limit.

Your instructor would study the answers you gave, and if necessary would explain in detail about how you can get training in motorway driving, or how to use the wipers etc.

The questions in the book are based on those used in the Associated Examining Board's Learner Driver Progress Checks which are used by many of the country's top driving instructors. You can help yourself prepare for any future extended theoretical driving examination by careful study of this book in conjunction with the AEB's question papers and your instructor. (Sample questions from the AEB list are shown on pages 138–9.)

1 Before driving any motor vehicle a driver must be in possession of

(a) A full or provisional driving licence for that vehicle

(b) A certificate of insurance

(c) An MoT vehicle test certificate

(d) A medical certificate from a doctor

(e) A driving test pass certificate

2 The minimum ages for driving motor vehicles are:

(a) 17 for driving a motor car (Category B)

(b) 16 for riding a motor cycle (Category A)

(c) 18 for driving a Heavy Goods vehicle (Category C)

(d) 18 for riding a moped (Category P)

(e) 21 for driving a Public Service Vehicle (Category C)

3 When learning to drive a motor car the law says you must

(a) Be accompanied by a person who has held a full licence for at least three years

(b) Not drive on a motorway until after you have passed your driving test

(c) Have a minimum number of lessons totalling 1½ hours for each year of your age

(d) Pass a written examination before you apply for the practical examination

(e) Sign your driving licence before you take the wheel.

Answers

1(a) This is correct; you must be in possession of a provisional or full driving licence for any vehicle you drive.

(b) A certificate of insurance is required that covers a specific vehicle, and identifies which drivers can drive. The actual certificate does not need to be in the drivers' possession however, but they may be required to produce it when demanded.

(c) MoT Car test certificates are only needed for vehicles more than three years old.

(d) You do not require a medical certificate in order to drive a motor car, although you do need to pass a medical examination to drive Heavy Goods or Public Service Vehicles. However, if you suffer from any illness or disability which may debar you from driving, or if you take prescribed medicines or drugs, you should consult your doctor and take his advice before driving. If your doctor is unable or unwilling to make a decision the Driver and Vehicle Licensing Agency at Swansea maintains a list of doctors who are willing to do this.

(e) In order to obtain a driving test pass certificate you need to get practice as a learner driver, therefore you cannot be in possession of a pass certificate during your lessons. Remember, when you have passed the test, to send your licence and the pass certificate to the DVLA at Swansea as soon as possible to have your licence changed.

2(a) The minimum age for driving a motor car is 17 unless the driver is in receipt of a Mobility allowance when they are allowed to begin learning to drive at the age of 16.

(b) The minimum age for riding a motorcycle is 16.

(c) The minimum age for driving an HGV is 21, unless the driver is in the armed forces or on a Junior Driver scheme through the Road Transport Industry Training Board.

(d) The minimum age for riding a moped is 16.

(e) The minimum age for driving a Public Service Vehicle is 21.

3(a) Anyone accompanying a provisional licence holder must be at least 21 years of age and have held a full licence for three years.

(b) Car learner drivers are not allowed to drive on motorways. Even if they have passed their driving test they should exchange it for a full licence.

(c) There is no legal requirement for any learner driver to have professional lessons. Nevertheless a good general guide to the number of lessons needed is indeed 1½ hours for each year of one's age.

(d) At the moment there is no legal requirement to pass a written theoretical examination, but all other member countries of the European Community except Britain and Ireland do have to. It is anticipated that the requirement will change in the next three years or so.

(e) It is a legal requirement that any driving licence must be signed. A new driver is not allowed to drive before a licence has been received and signed.

Part 2 Introduction to the controls and road procedure

At the end of this, the second stage of your driver training, you will be able to:

Identify and explain the use and operation of the main controls of the car;

Enter, start the engine, and move away from the kerb without any instruction;

Observe hazards while on the move;

Select all gears as required (including block changes) without looking;

Bring the car safely to rest, every time as required; and also quickly but with safety, as in an emergency;

Move away safely on any hill (up and down);

Move away at an angle, up to half a car's length from any obstruction;

Indicate safely (and signal by arm) while on the move.

Road-user behaviour

You are now going to learn to drive on roads where there is bound to be some other traffic. The earlier you adopt correct road-user behaviour, the better for all. There is no doubt about the most common cause of traffic accidents. The nut holding the wheel.

In over 95 per cent of accidents, driver or rider involvement is a substantial factor, and in 85 per cent of accidents, driver or rider error is the sole cause of the incident. Part of the problem over many years has been an acceptance of the word 'accident' to define the result of stupidity or carelessness on the part of road users. Another problem is that because people 'get away with' a silly or stupid action so many times, on the odd occasion that they are confronted with the result of their mistake they are tempted to put the blame on external circumstances instead of their own foolishness.

For example, if you turn right, and at the same time cut the corner of the road you are entering, you may get away with it. In fact you may probably get away with it four times out of five. On the fifth occasion someone may be parked round the corner and so create a confrontation situation. This means that you, or another road user, or both of you, will have to make a sudden decision. Provided your reactions, and those of the other road user, are correct and in time, the crisis is averted. But by taking chances, a small percentage of error will always remain. The moment you allow yourself to do something which is potentially dangerous you are putting yourself and others at risk.

From a road safety point of view it does not matter whether the mathematical probability is one in a hundred or one in ten. What is important is that every day fifteen people are killed on the roads in Britain, and that the error which you are about to make will be the one that turns into a confrontation, and that confrontation becomes a crisis. Road traffic accidents do not just happen. They are caused, mainly by driver error, and usually by two people who just happen to be careless at the same time and place. Your duty as a driver, both to yourself and to others, is to limit your own errors to an absolute minimum.

It is no coincidence that the marking system on the driving test is broken down into three stages of mistake. Simple faults, which do not involve any other road user and could not lead on to a confrontation situation, are marked, but are regarded as minor errors and have no effect on the driving test result. A fault which does involve other road users is marked as serious and results in failure; whilst a fault which causes another road user to take any form of avoiding action is marked as dangerous and obviously results in failure also. Translated into practical driving terms, cutting a corner which can be seen to be open and clear would be a minor fault. Cutting a blind corner, even if it happened to be clear, is marked as a serious error.

Cutting a corner when another road user or vehicle is involved would be considered dangerous.

The three basic causes of road traffic accidents are easily identifiable:

- following too closely behind another vehicle;
- turning right before knowing that the turn is safe; and
- overtaking and meeting other vehicles unsafely.

Unfortunately, driving too closely behind other traffic occurs so often that everyone takes it for granted. Perhaps only one incident in ten thousand results in a confrontation, and only one in five of these results in an incident. Experience, good forward observation and quick reactions all combine to prevent the likelihood of such an incident. However, even twenty years' experience and brilliant reactions will not save a situation where you need twenty or thirty metres to stop and the distance between vehicles is less than this. There are two ways to avoid becoming a road traffic statistic in relation to following distances. One is never to drive so closely behind another vehicle that you cannot stop in the distance you can see to be safe. The other is to plan your driving so that you are always aware of what changes to speed and position are possible by any of the road users ahead of you. Fortunately, one of the very first 'distances' you will learn as a driver is that of the eyesight requirement (this is 20.5 metres for a number plate.) This also happens to be the minimum distance at which it is possible to stop safely at 30 m.p.h. So a very simple rule to follow all the time you are having your early driving lessons is to remember that if you can read the number plate of the vehicle in front you may be too close. Drop back. If the vehicle behind is also too close to you for your comfort, give him the opportunity to overtake. If he does overtake you, then accept gracefully with the satisfaction of knowing that if it is his intention to run into the back of anyone it won't be you.

Avoiding danger when turning right is much simpler. You must never commence any turn, right or left, unless you are certain that you can complete the turn safely, without inconveniencing any other road user, including pedestrians who may wish to cross the road at that point.

The dangers of overtaking, meeting and crossing the paths of other vehicles are much more obvious. The essential thing is to give yourself a safety zone around your vehicle, and learn how to overtake and pass other traffic correctly and safely.

Always assume that all other road users are unaware of your presence and never take their actions for granted.

As well as the dangers to you from other road users and their actions, it is as well to remember that young children and older pedestrians are more at risk than other sections of the community. Although you are always in danger whenever you make a simple error at the same time as someone else makes a stupid one, when children and the elderly are around you need to take extra care. Children react quickly and often unpredictably. Older people often react slowly, and sometimes not at all.

The dangers of mixing drinking and driving are well known. No self-respecting driver these days would ever mix the two. However, it is easy to assume that other drivers and road users have the same regard to common sense. While it would be unfair to assume that a large proportion of pedestrian casualties are due in some way to their own intake of alcohol, you need to be extra careful whenever you are driving in the vicinity of public houses at any time. There are only two rules with regard to alcohol and driving: *never* mix the two yourself, and assume that everyone else does.

Drink, however, is not the only drug which affects drivers and driver behaviour. Even such things as aspirin, paracetamol and other medicinal drugs can cause drowsiness and seriously affect anyone's driving performance. Quite often a bottle of cough medicine will carry a warning not to drive. Always read labels before mixing driving with taking any medication. There are some drugs which are essential to certain people to enable them to live normal lives, those suffering from diabetes being a typical example. The results of not taking their prescribed drugs can result in being unsafe to drive, or make them need special consideration as pedestrians. It is essential that every driver, but especially the novice, takes every precaution to allow for the effect of drugs or medication on any other road user.

Illness, tiredness, fatigue and other health problems can also cause drivers to feel below par. Anything which makes you feel less than 100 per cent can also adversely affect the way you drive. A simple rule of thumb which you can always apply to your driving performance is that if you feel one degree under, drive five miles an hour slower. The effects of a bad night's sleep, overwork, worry and tension are all similar to being under the influence of a drug.

Your own personal health has a considerable effect on the way you drive. Pre-menstrual tension, pregnancy, colds, headache, toothache, even a row with a partner or colleague can adversely affect your driving. Insurance statistics show that most accidents to women, whether in the home or on the road, occur during the four or five days of their menstrual cycle. Being aware of this can make you more wary and careful. Men can also be seen to be more accident prone at certain times. One of the benefits of professional driving instruction is that a good instructor can often identify a client's natural cycles of awareness and sluggishness, which can vary according to the time of day or other variables. If you tell your instructor how you are feeling, it will enable the two of you to make the best use of your health while driving. Once more, it is not enough just to take your own health into account; you must bear in mind that those around you may also be suffering from medical, physical or mental pressures which affect their driving performance.

There is really one important message here: TAKE EVERYTHING INTO ACCOUNT WHEN YOU ARE DRIVING, OR LEARNING TO DRIVE. The three separate aspects which you need to study are:

- the vehicle you are controlling;
- your own person and the things which affect you; and
- other people, whether in charge of vehicles or not, and the mutual effect you will have upon each other.

Learning to drive has very little to do with learning to control a motor vehicle, much more to do with learning to control yourself, and a great deal to do with learning to control external situations. Vehicle control is what your preliminary driving lessons are all about. Situation control is what your driving test and future safety on the road are all about.

Perhaps the most valuable lesson you will ever learn is that courtesy and consideration for all other road users is not only the way to pass your driving test; it is also the best method of ensuring a long and uneventful driving career.

Objectives: Department of Transport Syllabus: Stage 2

You have now learned to:

- enter the car safely and carry out all safety checks;
- move away safely from the kerbside on the level;
- select the correct low gear when asked to do so;
- follow the road ahead, steering to follow the safety line;
- brake smoothly to slow down, and bring the vehicle to rest at a designated spot on the road.

The second stage merely builds on this. You must be able to:

> Identify and explain the use and operation
> of the main controls of the vehicle

Having learned to move off, steer, change gear and pull in safely, it may sound peculiar that you are only now expected to identify and explain the controls you are already using. In practice, of course, it is much easier for you to be able to use a control than to explain how it works. This is the purpose of this particular exercise.

Read, study and understand exactly what every control does (see pages 33–43) and then see if you can explain to your instructor what each does, how it does it, and – for an added bonus – identify correctly the words which he and you will be using every time it is referred to.

You will find the next objective easy too. After all, you have done it all before!

> Enter, start the engine, and move away from the kerb without
> any instruction
>
> Observe road hazards whilst on the move

Here, all you have to do is remember what you have learned in Stage One and show that you can put it into practice without help. Enter the car, carry out safety checks, start the engine and move off from rest without instruction.

When you have done that, pull in and stop at a predetermined spot you have identified to your instructor. Now do the same thing again, only this time you will drive further down the road, observing and taking the correct action for all that you see. Everything on the road ahead or to the side of you is a potential hazard; your skill is to learn which hazards require your immediate attention. (Reread 'Road User Behaviour', pages 57–61.) You are beginning to learn 'situation control'.

When you can attain this objective (remember, you must do it without any reminders or guidance) you are no longer a complete novice. You know *what* to do, but not necessarily *when*. The function of your instructor now is to take on the role of a co-driver. You will be using his brains and your hands and feet to drive the car safely and successfully. Every lesson you have now is aimed at getting you the maximum experience of all kinds of road and traffic conditions.

Able to select the correct gear

The correct hand positions for selecting gear

If you position your hand correctly you will always select the correct gear.

Find the neutral position (where the lever will move sideways against light pressure). Then remember that 1st, 3rd and 5th move towards the windscreen. 2nd and 4th move towards the seats.

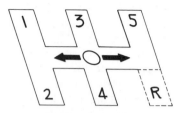

Make sure that the gear lever is in neutral
when the car is stationary

Place your hand so that it can press sideways and then up or down
for each of the gears. With a five speed gear box, 3rd and 4th will not
require any sideways pressure.

Block gear changes

Once upon a time gear boxes and the gears which meshed inside
them were built like tanks, and the brakes used to slow cars down
were weak and unreliable. Nowadays, brakes are superb, and gears
are flimsy affairs in comparison, so all drivers need to be trained to
use brakes for controlling the speed of the car and then to select the
correct gear for the next move. It is as pointless always to choose
each gear in turn when slowing down as it is to wear each item in
your wardrobe simply because it is there.

Block gear changing, as it is usually called, is just another way to
describe how you can select any one gear immediately after any
other. There is no need to go 'through the box', which is what the
old-fashioned way to change gear is called. Under normal
circumstances, you would always go through the box when changing
gear upwards (i.e. from 1st, to 2nd, 3rd, 4th and finally 5th). On the
way down, however, you do block gear changes. For example, you
might change directly from 4th to 1st when the vehicle is slowed
down, by the footbrake, almost to a stop.

The secret of learning correct gear changing is to be aware of which
gear you need, placing your hand correctly on the gear lever *before*
your instructor asks you to make the gear change. In your early
lessons, your instructor will have told you to prepare to select a
specific gear. Then, when you were ready to do so and the timing
was right, he would have instructed you to carry out the gear change.

Now you will be asked to select a gear without being told which one is needed. This requires considerable practice both with moving the gear lever correctly, recognizing the change in engine note that indicates a gear change is needed, and assessment of road and traffic conditions.

The correct selection of gears, without being told when and which ones to use, is the ultimate stage of this objective. Only when you can achieve this consistently, without error or instruction, are you ready for your practical driving test.

Bring the car safely to rest, every time as required; and also quickly but with safety, as in an emergency

The 'emergency stop' exercise is often thought of as a milestone in the training of any learner driver. Some instructors prefer to wait until just before the driving test before they teach their pupils how to do it because they regard it simply as a driving test manoeuvre. However, every instructor – and pupil – ought to be convinced that if a genuine emergency were to arise it would be coped with satisfactorily. Every pupil needs to recognize exactly how long it takes to stop a car in an emergency; and then to bear that distance firmly imprinted into their safety zone memory. Remember that it takes longer to stop when the roads are wet, and even longer with a loaded car. Remember, too, that this is one occasion when the clutch should be left completely alone until the car has almost stopped.

The practice emergency stops taught to you by your instructor will be almost leisurely affairs to start with. He will explain how it is done while you are stationary and then allow you to practise, first at 5 m.p.h., then at 10 m.p.h., then 15 m.p.h and eventually at whatever speeds he decides are necessary. On the driving test this will rarely be at speeds more than 20 m.p.h. In a genuine emergency, of course, there will be no such controlled conditions. It is therefore essential that your instructor ensures you know how to recognize the need for a real emergency stop, how to act in that eventuality and, most important of all, how to avoid emergency stops in the first place by being more aware of what is ahead of you and what may happen.

The skill of a novice driver shows itself in the way that he or she is able to slow down for a hazard without upsetting the balance of the vehicle. Imagine you have your grandma sitting in the back of your car at all times. She never looks out of the window and she is nursing a basket of eggs. Try to avoid upsetting her, and the eggs, at all times. Brake early and therefore smoothly and gently. If you do need to brake firmly, try to do so by braking progressively, becoming more and more positive until your speed drops to that which you need. Then ease off the braking so that your speed is adjusted gradually. Remember that if you brake firmly at 70 m.p.h. you may slow down to 60 m.p.h. with hardly anyone in the car noticing. If you are travelling at 10 m.p.h. and slow down to a stop everyone notices it. The faster you are travelling, the more firmly you need to apply the footbrake. The slower you are the more gentle you need to be with your right foot. But you need to avoid a scrambled grandma!

The sequence for slowing down or stopping is always the same:

- Look in your **mirrors** (including your door mirrors);
- Take your foot off the accelerator pedal as soon as you see the need to slow down. This will have a decelerating effect.
- Give an arm slowing down **signal** if you think it is important. If not you can rely upon the brake light signal to tell the traffic behind what you are doing. But don't frighten them.
- Put your foot gently on the brake and begin to press firmly.

Continue to brake progressively and positively until you are slow enough to cope with the hazard. If necessary, release the footbrake pressure just before you come to rest so that stopping is gentle.

If the intention is to stop, then put the clutch pedal down only a fraction of a second before the car comes to rest. That way you will avoid coasting, which is potentially dangerous and can lead to loss of steering control if you brake too harshly.

Your clutch control should now be good enough to satisfy the next objective easily, which is to be able to:

Move away safely on any hill (up and down)

Having learned to move off on the level, correctly and smoothly, now is the time to learn how to move off on a gradient. The hill start is probably the greatest single driving skill you will ever learn.

Once you know how to hold the car on the clutch you will be king or queen of all you survey. It is one of those skills which seems impossible, until you have mastered it, after which you wonder why you ever considered it a problem.

Operating the clutch

Try to visualize the clutch pedal as having a total movement of 'ten units'. It doesn't matter what the units are. Imagine that nought is at floor level, and 10 is fully up with your foot off the pedal. From 0–4 nothing happens, the car is still not able to move. From 5–10 the car is moving. Between 4 and 5 all hell breaks loose! At 4 and a bit the engine side of the clutch starts to meet the gearbox side of the clutch and begins to turn it (this is known as 'biting point'). If you haven't pressed the accelerator the engine will usually stall. If you have 'set the gas' just enough, the car will want to move forward slowly, unless you let the clutch up from 4 to 5 at this point when it will probably lurch forward instead. As a reaction to this lurching the inexperienced driver generally takes his foot off the gas pedal, then re-applies it, producing the phenomenon known as 'kangaroo petrol'. If, instead of this, the learner is able to hold both feet absolutely steady at this point, the car will be able to move slowly (against the handbrake). You now release the handbrake, still holding your feet steady, and the car purrs forward – very slowly. Now you raise your left foot slowly from 4.1 to 4.2, then 4.3, and continue through until 4.9 and 5; the car will be moving smoothly at the correct speed and you will have mastered 'clutch control'. Many learner drivers have problems because they try to use the whole length of their left leg instead of just the ankle joint. Imagine eating a boiled egg. Which is the easier to use, a tea spoon or a golf club?

The art of moving off downhill is even simpler. It requires the use of the footbrake instead of the handbrake and accelerator. Keep your right foot on the brake. Release the handbrake (parking brake). Allow your left foot *slowly* to bring up the clutch, and keep it still at the biting point. When you are ready to move away, take full observation, and move your right foot to the accelerator and slowly allow the clutch to come fully up.

The other two positions of the handbrake are: fully down, used when the car is moving; and fully applied, when you are stationary. Think of the three positions of the handbrake as: down when you are driving; up when you are parked; and held by your hand when you are waiting to move off.

Avoid ratcheting the handbrake when you apply it, which means making a clicking sound caused by not pressing the button in sufficiently first. Make sure when it is applied that your vehicle stays immovable. Check that the brake warning light goes out when you are moving. Never use the handbrake when you are on the move.

Once you can move away safely on any hill you will find it easy to manoeuvre your car in a confined space. The first exercise of this kind you will need to demonstrate is to be able to:

> Move away at an angle, up to half a car's length from any obstruction

Until such time as your clutch control skills have been perfected, moving away from the kerb will always have been 'interesting', in that you have never been sure how much room you have, or how much room you need. Now that you have complete clutch control you will find that this exercise becomes quite simple. You can demonstrate your skill by looking over your right shoulder as you actually move off to check your blind spot. The sign of the novice is that he still wants to watch the car ahead instead of checking the state of the road behind.

In Stage 2 you are required to be able to:

> Indicate safely (and signal by arm) whilst on the move

Two separate skills are involved here: the ability to recognize when a clear and unmistakable signal is required, and to give that signal – in good time – either by arm or by indicator, without interfering with steering and changing gear effectively.

Although all motor cars are now equipped with electrical indicators to warn others of your intention to turn right or left, there are occasions when an arm signal would be advisable as well (and indicators can malfunction). There is no great skill in this. You simply need to ensure that, when you remove your right hand from the steering wheel, you don't allow the left one to overcompensate the steering. It is important that you are able to steer correctly and safely with one hand while you are signalling or changing gear with the other. Practise your slowing down signal by arm whilst you are stationary before you try it on the move. Then see how easily you can signal with your right arm before slowing down, say, at a zebra crossing.

Pedestrian crossings

The most likely situation where your instructor will want you to think about the benefit of using an arm signal is when approaching uncontrolled pedestrian crossings where pedestrians are waiting to cross. The reason for this is obvious. Those ahead of you cannot see your brake lights and won't know for sure if you are stopping or not. If you give a clear and recognizable arm signal they know that you mean it and will be more prepared to start to cross when it is safe.

It is important that the signal you give is clear and correct. You should never wave people across the road; all you can safely do is clearly indicate your intention and allow them to make up their own minds. However, it is worth while bearing in mind that by giving a clear signal you help yourself in two ways: you encourage any pedestrians who are hovering on the edge to get across quickly and therefore save yourself having to wait longer than necessary; and you also ensure that anyone who might be liable to overtake you realizes that you are going to stop. This especially applies to motor cyclists and cyclists who might otherwise be tempted to try to get past you whilst they are unsighted from the crossing.

Purely as a matter of law, once a pedestrian has placed a foot on any part of an uncontrolled crossing that pedestrian has absolute priority. Therefore if you approach a crossing where someone is teetering on the edge, and signal your intention to allow them to cross (I am slowing down is all the signal actually says), you will encourage them to take up your offer. If you did not signal, though, and decided to drive on through, and they also decided to step onto the crossing before you had cleared it, you would be guilty of not giving them precedence. By signalling you can easily avoid that happening.

There are four rules you must know about uncontrolled pedestrian crossings:

1 It is illegal for a vehicle to wait or be parked within the area of the zig-zag lines.

2 It is illegal for any vehicle to overtake the leading vehicle in the left-hand lane on approach to the crossing in the zig-zag area.

3 It is illegal not to give precedence to any pedestrian who is actually on the crossing itself. (This means giving way to pedestrians who are crossing and suddenly decide to turn back too.)

4 Just to show that justice prevails, it is also illegal for a pedestrian to cross the zig-zag lines, or to loiter unnecessarily on the striped area.

Uncontrolled crossings become 'controlled' when they are governed by traffic lights or traffic controllers (a policeman or traffic warden who takes charge of both vehicles and users of the crossing). Pedestrians then only have priority when it is given to them by the controller.

Pedestrian crossings are usually zebras or pelicans, but also include those areas at traffic lights bounded by studs to enable pedestrians to cross in reasonable safety.

A controlled crossing

Achievement of objectives

As you draw to the end of the second part of this training book, you will need to confirm (to yourself) that you have achieved the objectives laid down at the commencement of this part. You can either complete this check sheet yourself as a personal commitment that you have reached an acceptable standard or you can invite your instructor to confirm your ability to perform each of the following tasks to the standard acceptable to him or her.

At the end this second part of the book I am now able to:

Identify and explain the use and operation of the main controls of the vehicle;

Enter, start the engine, and move away from the kerb without any instruction;

Observe road hazards while on the move;

Select all gears as required (including block changes) without looking;

Bring the car safely to rest, every time as required; and also quickly, but with safety, as in an emergency;

Move away safely on any hill (up and down);

Move away at an angle, up to half a car's length from any obstruction;

Indicate safely (and signal by arm) while on the move.

Signed:

Pupil: _____

Instructor: _____

Date: _____

Questions

Part 2

1 Before moving off from a parked position a driver should always carry out correct sequences of operation. These correct sequences of operation include:

(a) Handbrake applied, neutral gear before starting the engine

(b) Look through the mirrors, signal intention, pull out

(c) Signal intention, look all round, final check in mirrors

(d) Look in mirrors, check over shoulder, signal if necessary

(e) Look in mirrors, wait until road behind is clear, move off without a signal.

2 A driver should always apply the handbrake (parking brake) of a vehicle when

(a) Stopped for more than two seconds

(b) Stopped at traffic lights

(c) Waiting in traffic

(d) Parked at the kerbside

(e) An emergency stop is needed.

3 Road and Traffic signs are used to guide, warn or inform road users. New drivers can help to remember them by knowing

(a) All signs which must be obeyed are contained in circles

(b) All triangular signs are used to give warnings

(c) A blue sign says you must do whatever is contained in the sign

(d) Information signs are usually rectangular

(e) White lines on the road are not legally enforceable.

Answers

1(a) Handbrake checked, gear lever in neutral before starting is a correct sequence and must always be used each time the engine is started.

(b) Mirrors before signal before moving out is another correct sequence. However, if the mirrors are used correctly it may not be suitable to move out, and it may be possible to move off without using a signal if no one would benefit from one.

(c) Signalling before looking round or in the mirrors is wrong. Always check your mirrors, and if necessary all round as well, before deciding whether a signal is needed.

(d) Look in mirror – and act on what is seen – then check over your shoulder, then signal if necessary, is a correct sequence.

(e) Look in mirrors, wait until the road is clear, then move off without giving a signal can be correct, provided traffic situation allows this. Ideally you should never need a signal to move off, because no one should ever be around to benefit from the signal. Unfortunately ideal situations rarely occur.

2(a) There can be no hard and fast rule about applying the parking brake of a vehicle. Therefore it is incorrect to say that you should *always* apply the handbrake if stopped for more than two seconds or even longer. But a good guide is always to say to yourself, if you have time to ask 'should I apply the parking brake?' then you need to do it.

(b) The above rule also applies at traffic lights. If you have time to ask yourself you should. You should also apply the parking brake whenever you are on a hill.

(c) The answers to 2(a), (b) also apply when waiting in traffic. Use your footbrake initially both to stop the car and to wait momentarily, but always be prepared to use the parking brake to take over from the footbrake whenever it is necessary, safer or convenient.

(d) Whenever the vehicle is parked at the kerbside the parking brake should always be applied. There are some drivers who like to leave the vehicle in gear and don't bother too much about using the parking brake correctly. This is foolish. It is an offence, and stupid, to leave a parked vehicle without applying the parking brake correctly.

(e) This is one occasion when it is not only unnecessary, but downright unsafe, to use the handbrake. Never try to use the handbrake, or parking brake, when you are stopping in an emergency. If for any reason your footbrake is no longer working, then gently pulling the handbrake full on will slow the rear wheels somewhat. But even this can be dangerous. Both hands ought to stay on the steering wheel for any emergency.

3(a) All signs which are contained in circles must be obeyed, but be careful! There are also some signs which must be obeyed which are *not* contained in circles. The Stop sign, the Give Way sign, are two. Lots of others can be considered when you start to look at signs as a whole.

(b) Once again the question didn't mention the Give Way sign, which is also a triangle. Nevertheless in principle a triangular sign is used as a warning. This is why the Give Way sign is an upside down triangle, in order to make it different.

(c) Blue circular signs are indeed 'obligatory' or 'mandatory' and must be obeyed. However, some signs like 'Pass either side' can be viewed differently perhaps from 'One Way Ahead' or similar signs.

(d) Information signs are indeed normally rectangular and if you are looking for help when in a strange town the rectangular signs are the ones to look for.

(e) White lines on the road are indeed enforceable, especially those which say give way, stop, or are part of the double white line system.

Part 3 Development of car control and situation control

At the end of this section, covering stages three to seven of the Department of Transport's syllabus, you will be able to:

Recognize and cope with each of the following:
Bend; corner; turning; junction; crossroad; roundabout (Mini, normal, large and unusual);

Negotiate with each of the above, safely and correctly:
on approach, while on them and on leaving;

Recognize, identify, and cope with all types of road and traffic signs, while driving in light traffic conditions;

Identify motorway signs, and explain their meanings;

Show a high standard of knowledge of the highway code and other motoring matters;

Reverse in a straight line;

Reverse round a corner to the left, on level ground;

Turn the vehicle in the road, using forward and reverse gears, safely and correctly, and with due regard to all other road users;

Turn the vehicle in the road, accurately with regard to various widths of the road;

Reverse round a corner to the right, on level ground;

Reverse to the left and right, while on all types of gradients;

Turn the vehicle round in a confined space (such as a car park) correctly and safely, under full control, correctly positioned and with maximum observation and care for all other road users; and

> Park the vehicle, using reverse gear, in any suitable confined space, between other vehicles, and at a parking meter bay, both on the right and the left.

Introduction

As your driving lessons progress you will find that they seem to alternate between periods when you are learning lots of new skills and when you seem to be stagnating without learning anything new at all. This isn't so of course; you are simply passing through what is referred to as a learning plateau, where you are consolidating what you have already learned. What you need to do is to use all your training sessions to your best advantage. By now all your control skills should have been learned quite well, even if you have occasional lapses. Ideally, you should be able to select each gear as necessary without being told which one. If you still need to be told 'when' to change, this is where you can concentrate your efforts. Listen for the engine noise, and test yourself to see if you can start to recognize when your instructor is going to say 'now'.

After a while, especially if your training routes have been planned on the basis of repetition, you will identify a situation which needs a gear change before you are told to do so. This is part of the process of learning to look ahead. As a pedestrian you will have been used to looking some five or ten **feet** ahead of where you are walking. As a driver you need to look five or ten **seconds** ahead of where you are now. At 30 m.p.h. this means over 250 feet or 75 metres ahead, though the actual distance you look and plan depends to a great extent on the traffic and road conditions. But perhaps the best way to learn how to anticipate is to get your instructor to give a running commentary in the early stages and allow you to try to do the same as your skill and experience grow.

Know your vehicle

The Construction and Use Regulations sounds a fairly boring subject, and unless you are a vehicle mechanic they do not make very enduring or enterprising bedtime reading. Nevertheless, as we continue to point out, the driver of every vehicle is responsible for the condition of that vehicle, not only in law, but morally too. While you are having professional driving lessons with a professional Approved Driving Instructor you can be certain that he will make sure the vehicle you are driving complies with the law in every respect.

However, it is not enough just to take his word for it; during your early lessons you need to have quite a few 'mechanical principles' explained in detail to you. Take a pride in the vehicle you are driving on your driving lessons and you are less likely to have to suffer from any mechanical breakdowns, or possible prosecutions, when you have passed your test and possess your own car later.

One of the best ways to learn your responsibilities is to encourage your instructor to detail what he does, each day, each week and each month, to make sure his vehicle complies with the law in every respect. If your driving lessons all take place with a car that is always warm you could be surprised the first day you drive your own car to find out that you have to use the choke to get the car started. Make sure you know what daily and weekly checks are required and how to carry them out. When your instructor arrives with the car for your lessons, ask what checks he carried out, and which ones could you take responsibility for. Before you get into the car at all, walk all round and look at the tyres. In subsequent lessons get your instructor to allow you to check the pressures and to compare them with those recommended by the manufacturer.

Certainly as a learner driver you should be able to check that the windscreen and all windows are clean and free from obstructions before every lesson. It is important to know that the only items which are allowed to be stuck onto the glass of your windscreen are the Tax disc, the green ADI certificate of your driving instructor (unless it is a pink one to denote he is a trainee) and the mirror. All other items should be removed from the glass as should anything which may cause reflections. Check the windscreen washer bottles have enough fluid in them, and that all the lights are in good working order.

Finally, before you are ready to begin your lesson you should make sure that everything on the seats, or loose in the car, is secured properly.

If you've brought a shopping bag or something similar with you, make sure that nothing can fall out and roll about or get wedged underneath the pedals.

Ask your instructor to earmark time in a later lesson for you to learn how to make checks on the water in the radiator, battery levels, and the various other regular checks that will be described in detail in the manufacturer's handbook. When you have passed your driving test and bought your own car for the first time remember, too, to call your driving instructor back again for another lesson on all you should know and do as the responsible owner of a motor vehicle.

By now you will know all the controls of the car you are driving; nevertheless knowing which is the wiper switch and which is the indicator will not prevent you from making the odd mistake every now and again, and you really ought to be completely certain that when the rain starts to fall you can operate the wipers and washers immediately and correctly. As your lessons progress you will be able to do these without your instructor having to tell you at all. Similarly you will have to learn how to operate the various heater and demister controls.

Road procedure and situation control

In Stages 3 to 6 of your training, you will be concentrating on the negotiation of bends, junctions and roundabouts of various kinds, and will probably be introduced to the art of reversing the car. These manoeuvres all require *perfect clutch control* – arguably the greatest practical skill you are going to learn (see page 67).

In these early days you may well feel that you will never master this skill. Be patient. Your instructor will have a specific hill in your area that he uses as a yardstick. When you can move off safely and correctly on that particular hill your clutch control is perfect. But until that day comes you are likely to feel less than completely confident.

Bear in mind there are no great problems in learning to drive, just challenges. Treat each challenge as it arises and see how best you can conquer it. Remember too that your instructor will so plan your lessons that these challenges are part of your structured training programme. Each time you overcome one challenge you are ready to move on to the next, and the next. Life, and driving, are full of challenges. Without them, learning would be dull and uneventful. Your instructor's role is not just to sit in with you and supervise what you do while you are learning to drive. He will be actively planning each of your lessons to make sure your learning is programmed, progressive and productive.

Once you have achieved good clutch control you can now begin to use it to your advantage in your approach to correct road procedure. Road procedure is what good safe driving is all about. Once you understand the basic principles of where you stand in the pecking order of driving sequences you will more readily take your priority when it is offered to you. In fact unless you take your priority when it is offered and available, you may well find you are creating more problems than you should.

Turning right and turning left are skills which present no real problem. The sequence you need to follow is easy to see and predict. Straight on at cross roads is also easy once you recognize whether you have priority or not. The only times you are likely to find confusion is at unusual roundabouts which do not follow the normal pattern of turning left and giving way on the right; this is where the advice and guidance of your instructor are absolutely essential. Listen carefully to what you are told and make sure you have sufficient confidence and faith to put it into practice. What you are learning at this stage is the beginning of 'situation control'.

The lessons which need to be learned at this point are much more concerned with fitting into road and traffic situations which exist all around you. Try not to think of yourself as a learner driver attempting to watch what happens from afar, but realize that you are part of the whole traffic pattern, and that your rights and sequences must be taken correctly. In this way you will learn a safe pattern of driving which will stand you in good stead for the rest of your driving life. When you are convinced that you are part of the traffic pattern you will find it easier to make decisions, not only in correct sequence but at the correct time too. Driving is very much a thinking game; those who are steady and cautious but who are aware of what is happening around them will be safer than many drivers who have experience, but have forgotten to concentrate on what they are doing.

Driving requires relaxed concentration. When you have gained experience you will find that to a certain extent some of your actions, such as selecting gears, braking, steering round a bend, can all be done without conscious thought. Some people refer to this as driving on autopilot and it won't take long for you to develop this particular skill. If you need a comparison, think in terms of eating with a knife and fork. No problem; but when you first started you probably shoved your food all over your face! Similarly, when you first start to drive you wonder how you could ever manage to change gear easily and subconsciously. Yet after a few lessons, some of the things you found almost impossible have now become second nature to you. Even with experience, most of us occasionally panic – say at a posh banquet with a great array of cutlery. Driving is the same: every now and again you have to come out of your subconscious in order to concentrate on something special.

What you really need to achieve during these early lessons is to be able to think and plan ahead for yourself. This is the real art of driving. By planning your route and progress through assorted road and traffic conditions, other traffic will no longer prove to be a problem or a danger to you, because you have taken their actions into account beforehand.

You need to consider all the separate driving tasks as a single entity. 'To follow the road ahead at all times with due consideration for all other road users and traffic situations.' If you can do this you can drive safely, and what is more a driving examiner would be willing to sign his name to this effect.

Not all of the driving tasks are concerned with your controlling a moving vehicle. You are also responsible for your actions when you are parked. The rules on parking are quite simply based on sensible relationships with all other road users. If you park badly, say with two wheels on the pavement, you will not only inconvenience pedestrians, who have every right to be there, you will also be breaking the law. The whole of the Highway Code is based on common sense and courtesy to others. If you are ever in doubt, why not imagine how you would like others to behave to you, and you can then do likewise.

MSM/PSL/LAD

You can summarize your responsibilities to other road users in a few brief series of sequences, the most notable of which are:

M.S.M. Mirror, Signal and Manoeuvre;

P.S.L. Position, Speed and Look; and

L.A.D. Look, Assess and Decide.

The way in which these work together can best be demonstrated by your approach to any junction when it is your intention to turn to the left, to the right or to continue straight on. The actual decision doesn't matter at this stage. All you have to do is to apply the above three sequences, in order, and all the decisions you make become simple and positive ones.

Mirrors always, before you make any change whatsoever;

Signal either by arm or indicator, whichever is appropriate;

Manoeuvre.

Now you can commence the next stage which is:

Position take up the correct position for the action to be taken;

Speed adjust your speed (usually by slowing down) to suit your intended action; and then

Look look all round and actively seek out those whom you may involve in any way.

Looking actually consists of three phases:

Look if you are approaching a junction usually look right first;

Assess next look to the left and begin to assess the situation;

Decide you can only decide what to do when you are in full possession of all the facts. Looking and then assessing are the only ways in which this can be done.

This sequence of road procedure must be applied every time that you approach a turning, junction or crossroad. Initially it may seem quite a long and complicated procedure, but when broken down into a practical application it can be seen to present a logical sequence.

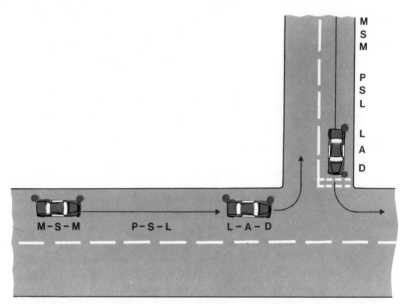

The sequences are always the same when turning left:
mirrors, signal, manoeuvre – position, speed, look – look, assess, decide

Not only is it logical, it is also the safest possible way to negotiate meeting other streams of traffic. If you think of a manoeuvre as being any likelihood of changes of speed or direction, then it is essential that a proper recognized signal must be given to anyone who may be interested or affected. The only way to know who may be involved is to make a careful check of mirrors and all other forms of observation. When joining other traffic at narrow angles it is often necessary to look over your left or right shoulder to be able to take them into full account.

It is worth while mentioning here the importance of eye contact with all other road users. It is this lack of eye contact between two road users which can lead to incidents. Although 85 per cent of all road accidents are solely caused by driver and rider behaviour most of them could be avoided by looking for and making eye contact with any other person before committing yourself to a particular course of action.

People very rarely have accidents because they want to, nor do they deliberately commit a stupid act. Most road traffic accidents happen because road users fail to carry out simple, safe sequences. It is for this reason that it is so important to learn correct road procedures right from the beginning of your lessons and that you put them into practice every time. It is also just as important that you have chosen a professional driving instructor who will spend sufficient time teaching you the correct attitudes towards driving and other road users. The cost of even the simplest road traffic accident is much greater than any money spent on professional driver training.

Whenever you find your driving lessons seem to be in the doldrums, go back to the earlier stages and see what you have achieved. Read the objectives at the beginning of these various stages of learning to drive. See how many of them you have already mastered. Then look forward to using these skills to conquer the next few remaining ones.

Check your personal record progress sheet in the beginning of the book and see just how far you have succeeded. And also have another read of the questions at the end of each part of the book. If you know the answers well, why not ask the questions of some of your friends or family? Perhaps the fact that you might be able to teach them something might boost your confidence just a bit.

Objectives: Department of Transport syllabus Stage 3

You are now driving in all sorts of road and traffic conditions, continually rehearsing those aspects of 'road procedure' you have learned so far. It is now time for you to learn how to recognize and define the various types of road formation, and to be able to demonstrate your ability to cope with them in practice. (See page 95 for information on road markings and signs relating to this topic.) You must be able to:

Recognize and cope with each of the following: bend, corner, turning, junction, crossroad, roundabout (including mini, normal, large and unusual)

- A **bend** is simply when the road you are already driving along changes direction

- A **corner** is a bend where one has the option of continuing along the road you are on or turning off on to another

- A **turning** is a corner where you enter a side road from a main road

- A **junction** is a corner where you turn from a side road into a main road at which you must turn left or right (you cannot go straight on)

- A **crossroad** is where two roads meet with an option to turn left, turn right, or go straight across (in other words, a multiple junction). Priority is usually given to one set of traffic, either permanently or by traffic controllers.

- A **roundabout** – no matter what its size – is in effect a crossroad which has been given a one-way system to aid traffic flow. Roundabouts are one-way streets which you enter by a *junction* and leave by a *turning*. The intention is to allow the traffic to flow more easily than it would via any form of controlled junction. Roundabouts can vary in size from simple Y junctions to complicated gyratory systems.

In order to demonstrate your total ability to drive competently, it is essential that you are able to:

Negotiate each of the above, safely and correctly: on approach, while on them, and on leaving

Junctions and crossroads

The procedure for turning into and out from roads, whether to the right or left, always follows the same basic rules:
The M–S–M (page 81) routine must be used. Once you have worked out what sort of junction it is, and where you intend to go, start your M–S–M sequence in plenty of time.
The P–S–L routine (page 81) must also be used. The second M of M–S–M stands for Manoeuvre, and in the case of turning a corner, P–S–L is the manoeuvre.

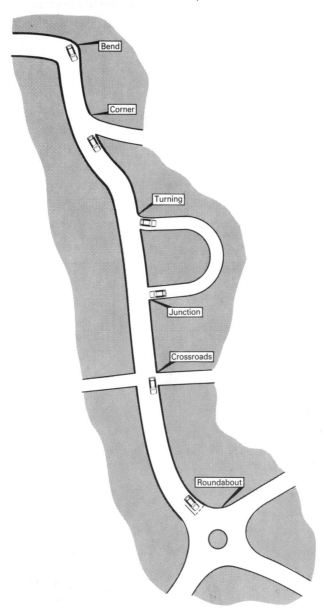

You must be able to recognize and cope with
these road formations

Right and left turns: a turning sequence

Car 1

When turning left into the side road, car 1 should normally be able to make his turn without interference, provided there is no obstruction on the road he is turning into, and he gives way to any pedestrians who might be crossing.

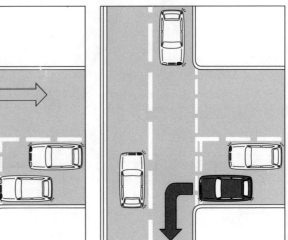

Car 2

Meanwhile, car 2, turning out of the side road, will need to make sure that there is no traffic on his right and that the road into which he is turning is also safe. Therefore, he needs to look to the right, left and right again before deciding to stop or continue. If he stops, it should be where he can still see but is safe.

Car 3

Car 3 is turning right into the opposite side road. He would ideally like to wait at the intersection of the two white lines but he is willing to be guided by traffic conditions. This position would enable him to complete his turn whenever a safe opportunity presents itself.

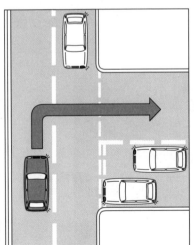

Car 4

Car 4 must wait until everyone else has been able to go before he emerges into the main road.

Thus, the turning sequence is car 1, then 2, 3 and 4.

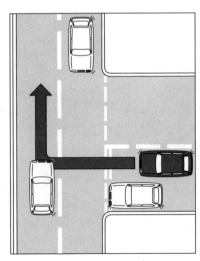

This means that after you have used your mirrors to determine who will be affected by your action, and signalled by indicator your intention to turn, you should now take up the correct position in the road. If you are turning left, you continue to follow the safety line. If you are turning right, then you should (using your mirrors again) move out to the turn-right position. Only then do you adjust your speed by use of decelerating or braking.

The final sequence is L–A–D– routine (page 81). You **look** to the right, normally, because this is where the initial danger lies. This first look merely shows you what may be approaching. Your next **look** is to the left. This gives you a chance to **assess** the situation. If necessary you continue to look right and left until you are able to assess the situation completely. Only then do you make the final looks to the right and left to make a **decision.** The decision – which must always be made **before** you cross any give-way line or merge into any other traffic – is either to **stop** or **go.** If there is any doubt you will always stop. If you are convinced that it is safe for you to enter or emerge into that turning or junction then you should do so. You must then check your mirrors and continue on the new safety line, picking up speed to join the new traffic flow.

'Cutting corners' is something we all do, but to cut corners in a car is potentially dangerous. In the illustration below the driver cuts a corner with no other traffic around – apparently. But what if he had failed to see an oncoming car? Cutting corners can easily become a habit, so don't do it. Ideally you should turn every corner as if there

Cutting corners like this can easily become a bad habit

were traffic coming out. Pretend there is a flag pole sticking out where the white lines intersect and you will always be safe.

Needless to say, you will have practised all of the various options of turning to the left and right, emerging to the left and right, as well as going straight across major and minor crossroads. One note of warning: it is worth while to remember that, even though you appear to have priority, someone coming the other way may fail to notice this or be unable to stop for you. Always **look** and be prepared to give way rather than insist upon your rights and risk causing an accident.

Turn every corner as if there were traffic coming out

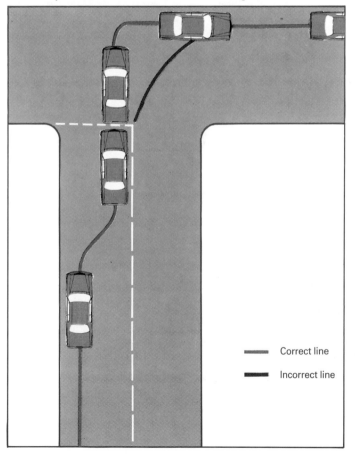

Correct line

Incorrect line

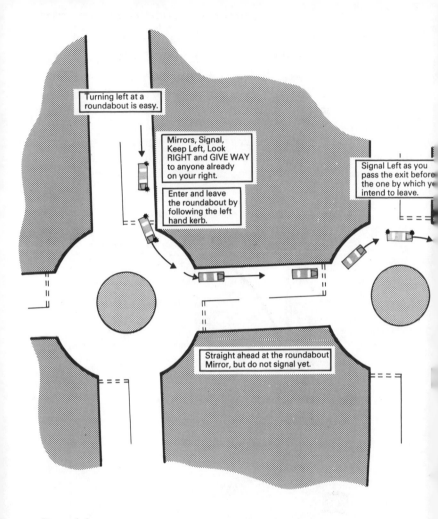

Turning left at a roundabout is easy.

Mirrors, Signal, Keep Left, Look RIGHT and GIVE WAY to anyone already on your right.

Enter and leave the roundabout by following the left hand kerb.

Signal Left as you pass the exit before the one by which y[e] intend to leave.

Straight ahead at the roundabout Mirror, but do not signal yet.

Roundabouts

The purpose of roundabouts is to change what might otherwise be an awkward crossroad or junction into a system which allows for smoother flowing traffic. At almost all crossroads and junctions these days priority is established, either by white lines or traffic lights. This means that certain flows of traffic can always take precedence over others. In some cases this can lead to hold-ups in the traffic flow from some directions. By changing the physical shape of the junction into a roundabout system, all lanes of traffic should have an equal

The correct way to
use a roundabout

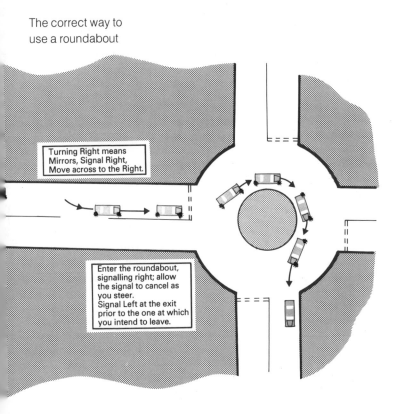

Turning Right means
Mirrors, Signal Right,
Move across to the Right.

Enter the roundabout,
signalling right; allow
the signal to cancel as
you steer.
Signal Left at the exit
prior to the one at which
you intend to leave.

opportunity to flow, and each vehicle is normally expected to give
way to traffic which is on its immediate right.

Roundabouts come in all shapes and sizes, from mini-roundabouts
with a simple painted dot in the centre to huge one-way gyratory
systems which sometimes take in a whole town centre. However, the
principle of traffic flow is always the same, and the most obvious,
and common, sort of roundabout is the simple crossing of two major
roads. As the latter is probably the easiest one to begin with, your
instructor will no doubt include one fairly early in your training route.

Traffic on a roundabout always flows clockwise; it always keeps left, and the dangers always appear on the right. There may be exceptions to these rules, but where they exist the markings are made exceptionally clear. Because you always give way to the traffic immediately on your right the only reminder of this is the single broken white line as you enter the roundabout. It is emphasized that the purpose of all roundabouts is to allow for the free flow of traffic round them, therefore the skill which your instructor will be teaching you is to approach, enter, drive round, and leave each roundabout without needing to stop unnecessarily. Unlike junctions and traffic lights, the priority taken at roundabouts depends on the sequence and arrival of traffic. Thus even though you are a new driver, you should take your priority when it comes, or the driver behind you may get impatient. The essential sequence must begin with approaching the roundabout at the correct speed and in the most suitable gear. You should be slow enough to merge into any traffic on your right if it is possible. On approach you will be concerned with full observation. That means keeping a safe distance from the vehicles in front of you, looking to the right to assess which gap you intend to enter, and also taking into account any traffic which is following you, so that you don't stop suddenly and take them by surprise.

Turning left

In your early lessons on roundabouts your instructor will teach you to turn left. This is the simplest manoeuvre and allows you to get the feel of how roundabouts work without getting too involved.

The sequence for taking the first exit on a roundabout is simple. It is exactly the same as if you were turning left into a major road. Your speed on approach is such that you can stop safely before you cross the give-way line if you need to. Your observation is divided between looking ahead and where you want to go; at the traffic already on the roundabout and who have priority over you; and in your mirrors at traffic following. An extra precaution at this stage is to remember that some cyclists might be tempted to sneak through on your left even though it is their intention to go round the roundabout, so take an extra look in your left door mirror. Entering and leaving the roundabout is simple. If it is clear on your right, and you will not inconvenience anyone at all, continue round with a left indicator flashing all the time, then leave the roundabout by the first exit, keeping in the left lane all the time.

Going straight across

If you are going forward at the roundabout the approach sequence is the same as for turning left. Again, you will normally use the left lane. Do not signal on approach to the roundabout, but do use your left indicator as soon as you are level with the exit immediately before the one you wish to take. When the roundabout has two lanes on approach, it is possible to follow a road straight ahead by keeping in the right-hand lane round the roundabout, but this is something that your instructor will only teach you when circumstances allow and you are much more experienced.

Turning right

When you intend to turn right at a roundabout, you should approach in the right-hand lane, and keep to that lane as you arrive and go round the roundabout. You will signal right on approach, keep your right indicator flashing as you enter and go round the roundabout, changing it to a left indicator at the exit *before* the one you intend to leave. You would normally leave the roundabout in the left-hand lane.

Your instructor will always brief you precisely about which lane to use on each roundabout you encounter. Fortunately, the system of lane markings at roundabouts is now very good, and almost all multi-lane roundabouts are now well signed. Your instructor will always tell you which lanes to use for entering, travelling round and leaving. As you get more skilful you will be able to make your own decisions based on where you are going and what lane markings exist.

Some of the more fancy roundabouts can appear confusing at first, especially what are euphemistically called 'Magic Roundabouts' where the original very large gyratory system with an enormous centre section has now been transformed into a series of six or seven mini-roundabouts, where you apparently travel round the large roundabout in the wrong direction. The secret is to forget the large roundabout completely and treat each mini-roundabout as a completely separate exercise.

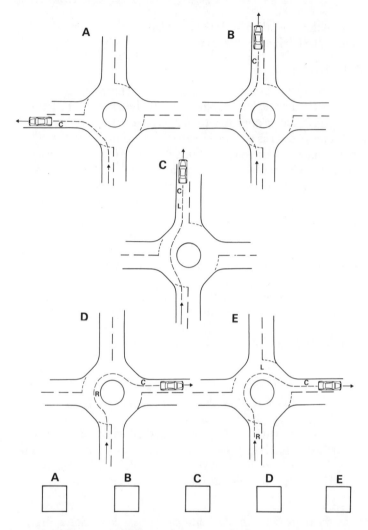

In each diagram the dotted lines and arrows show the route taken at a large roundabout. The letter L means left signal on. The letter R means right signal on. The letter C means cancel signal. In which of these diagrams is the signalling correct?

Mini-roundabouts are simple to cope with. Generally speaking, they consist of a Y or T junction with a painted or concrete blob in the middle. Priority is shown and you give way to traffic immediately on your right. The sign on approach is usually one showing three white arrows forming a circle. Ideally you should approach, enter and leave them exactly as you would a conventional roundabout. However, not all road users respect such roundabouts, and you may encounter an oncoming driver who will ignore the painted or concrete centre completely; bear in mind, too, that drivers of large vehicles, especially articulated lorries, often have to take the only path possible which may well mean crossing your road space. Always be prepared to give way to anyone who may need more consideration, or who has already occupied 'your' space.

Remember that roundabouts are intended to help traffic flow. The basic sequence of M–S–M , and P–S–L always applies, and as a learner driver you are trying to learn a skill which will enable you to cope with any traffic situation anywhere. If another road user breaks the rules, don't try to punish him for it. Allow him to get out of your way, and then you can continue driving and coping safely.

Objectives: Department of Transport Syllabus Stage 4

As well as coping with each of the above types of road junctions, it is essential that you are also able to:

Recognize, identify and cope with all types of road and traffic signs, whilst driving in light traffic conditions

and identify motorway signs and explain their meanings

Perhaps the most obvious aids to forward planning and anticipation are signs and road markings. Every learner driver must learn the meaning and purpose of all such signs.

Signs and signals

Signs come under three basic headings:

- what you *must* do or *cannot* do
- what to look out for (warnings)
- information

Some of the different types of road sign

Round signs are easy to identify and the most important ones to look out for because they give you orders. If they are *blue* circles they are telling you that you *must do something*. If they are *red*, they tell you *what you must not do*.

If you see a triangle, then it is a *warning*. The single exception to this rule is the triangular 'Give Way' sign, and this is easy to identify because it is upside down and much larger than other triangular signs.

The final group of signs are the rectangular ones. They are also important as they give you *information*.

- Circles give orders
- Triangles warn
- Rectangles inform

Some of the signs you come across in the books will not be ones you will meet as a learner, for they are motorway signs.

 Maximum speed

 Move to lane on your left

 Leave motorway at next exit

 Do not go beyond this signal

 Restriction is ended

Overhead motorway signals

Nevertheless, at this stage in the learning process you should begin to familiarize yourself with them, so that you understand what they mean and where you are likely to find them. Some motorway signs are found in the central reservation between the two sets of carriageways, others are seen overhead. Some signs are permanent, others are temporary and changeable because they are spelled out from a central control office. Some motorway-type signs are also found on trunk roads. Although you are not allowed to drive on motorways yet, it is still helpful if your instructor can devise a few routes which cover motorway type roads, perhaps local trunk roads or town bypasses. Your practical driver training should have reached the stage by now of needing as much variety in the choice of training routes as possible. Get your instructor to show you where these roads are, and if possible drive on them.

Traffic lights, lines and other traffic controllers

Traffic controllers are there for the purpose of easing traffic flow; wherever roads join or cross each other some means has to be found to allow them to mingle without danger. The most common way this is done is by 'Give Way' or 'Stop' lines and signs; it then becomes clear which traffic has priority. In order to improve traffic flow, the priority can be changed so that alternate roads have it: traffic lights are the most effective method for achieving this. Many sets of lights also incorporate a pedestrian phase and in some towns, where there is a high proportion of cyclists, there may be allowance for them to go slightly ahead of cars.

Not all road and traffic signs are placed on poles. Some of them are painted on the road. The 'rules' for white lines are simple; the more paint used, the more important the instruction. A single broken line down the centre of the road denotes a simple hazard warning; this is usually at a cross road or junction, or perhaps on a bend or bridge. If the road bends a lot then use is made of double white lines which separate lanes of opposing traffic. If these lines are both solid, no one may cross in either direction. If one line is solid and the other broken, you are only allowed to cross when the broken line is on your side, and then only if it is absolutely safe to do so.

Yellow box junctions are painted in some parts of the road to show where you should not wait. They especially apply when a right turn is involved and are intended to prevent hold-ups. If common sense

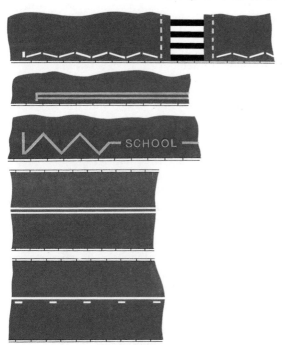

Some road and traffic signs are painted on the road

were applied by all drivers they would not be necessary; but it is an offence to disobey them so it is important to remember how they work. You must not stop in any area covered by yellow criss-cross lines, unless you are turning right.

Stopping on yellow box junctions is only permitted when you are turning right

In this case you may enter and wait in the box provided that you can see that your way will be clear to make the turn behind any oncoming traffic. If there are already two or more vehicles in the box waiting to turn, it's best not to enter.

As your driving lessons progress, make an effort to understand every sign you come across. And when you have passed your test and continue to drive on your own, take a pride in never missing a single sign.

Road and traffic signs are not just there to be noticed and identified. The real purpose is for them to be obeyed where this is possible. Some of them, like speed limits, or indeed any red or blue circle, *must* be obeyed. But even warning signs often need you to take some form of effective action. Many learner drivers completely ignore such things as the maximum height of vehicles under bridges, knowing full well that every bridge must obviously allow passage for a family saloon car. However, after they have passed their tests they may well become conditioned to ignoring such signs, only to discover their purpose the hard way the first time they try to take a large or highsided truck underneath a low bridge!

It is also worth remembering that all speed limits are covered by the appropriate sign. If you are in a built-up area, that is a road where there are street lamps at intervals closer than 200 metres, then a 30 m.p.h. limit exists. If not, then repeater signs (also at 200-metre intervals) will tell you what the speed limit is. If there are no signs at all then the speed limit is 70 m.p.h. for dual carriageways and 60 m.p.h. for single carriageways. So the existence or absence of street lamps and speed limit signs will always tell you what the speed limit is on any road.

- No lamp posts, no speed limit signs: dual   = 70 m.p.h.
  carriageway
- No lamp posts, no speed limit signs: single   = 60 m.p.h.
  carriageway
- No lamp posts, speed limit repeater signs: all roads   = as sign
- Street lamps, no speed limit signs: all kinds of road   = 30 m.p.h.
- Street lamps, repeater signs, 40, 50, 60.   = as sign

One extra item on the use of signs and signals is the fact that some people use unofficial signs, such as the flashing of headlamps, or occasionally the waving of an arm, when they are trying to tell you something. There are obvious dangers in using these signals. For instance, some drivers may well flash their headlights briefly at you to say you can come on through a gap; others may flash theirs for a slightly longer period to let you know that they feel inclined to take priority even though it may be in doubt. Your problem is to decide which signal is intended, and act accordingly. The problem is exacerbated by the fact that one driver's idea of what constitutes a short flash is another driver's idea of a long one. Therefore you can only ever play it safe.

As well as understanding the meaning and purpose of all signs and signals, you must be able to

> Show a high standard of knowledge of the Highway Code and other motoring matters

This means that you are not only required to learn what the Highway Code and other books of this nature contain (e.g. *Know Your Traffic Signs*, HMSO), you must also be able to put their principles into practice at all times when you are driving.

Objectives: Department of Transport Syllabus Stage 5

If you examine the objectives for Stages Five to Seven you will see that they cover the subject of manoeuvring under total control in great detail. The major part of your practical training is now concerned with clutch and vehicle control.

The Fifth Stage of your training requires that you are now able to:

> Reverse in a straight line

When is the best time to start learning to reverse a motor car?

There are two points of view: one says that the best time to learn anything is when you ask questions about it, because the interest and motivation are there. The second answer is probably better; when you have perfect clutch control over moving off forwards. If you can make the car move slowly forwards it is not difficult to make it move slowly in reverse. All the manoeuvring exercises you are likely to do in a motor car are concerned with parking or changing direction. Therefore you will always be doing them slowly and with full regard to every other road user.

For straight-line reversing, your only concern is to take full observation to the rear, which may well involve turning in your seat. The secret is to aim straight. Look at where you want to go – not at your instructor. Ideally you should be aiming at the vehicle behind you, and if you keep looking at it you won't be tempted to steer incorrectly. Drive slowly, under clutch control, keeping your eyes firmly on the target area.

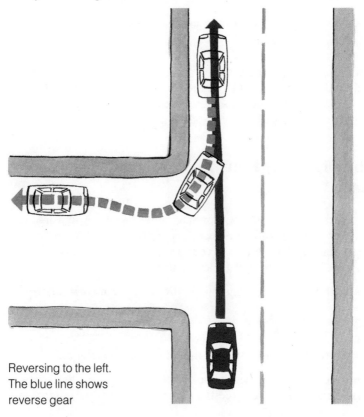

Reversing to the left.
The blue line shows
reverse gear

Once you can reverse in a straight line, it is then reasonably simple to add to that ability a demonstration of how you are now able to

Reverse around a corner to the left, on level ground

Reversing around a corner consists of three stages (that magic 'three' again): reversing in a straight line, turning the corner, and reversing in a straight line again. That is why it is important to be able to reverse in a straight line first. There are three skills involved in this exercise: the ability to make the car move very slowly in reverse gear; the ability to turn the wheel very gently and smoothly; and the ability to look all around you to ensure that no one else and nothing else can be inconvenienced by your manoeuvre.

The same three skills are also involved in the next exercise, usually called a three-point turn. For those of you who like to think of it as such, the three points again are:

● absolute clutch control;
● precise steering;
● full observation.

You must be able to:

Turn the vehicle in the road, using forward and reverse gears, safely and correctly, and with due regard to all other road users

Initially this needs to be done in any reasonably wide road, with an adequate camber. Later on in your lessons your instructor will become a little bit more demanding and use this turning-in-the-road exercise to test your control and observational skills to the limit. You must be able to:

Turn the vehicle in the road, accurately with regard to various widths of the road

There is of course no requirement to turn the car round in three movements; this is why it is not called a three-point turn by examiners and instructors. However, if you associate the three points with what you have to do to demonstrate the manoeuvre successfully, it doesn't matter if you think of it as a three-point turn.

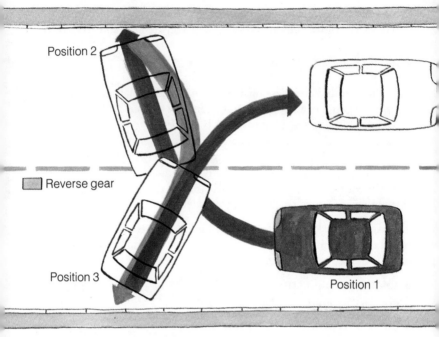

Position 2

Reverse gear

Position 3

Position 1

The turn in the road. Note that it is *not* essential
to complete this manoeuvre in three turns.

Once you can control the clutch on a hill going forward, you have to
use the same skills to make it go slowly in reverse, and at the same
time concentrate on turning the wheel from lock to lock. However,
neither of these two skills is as important as the matter of looking all
around you to make sure that no one else is being inconvenienced.

So there are three points to remember about the turn in the road:
clutch control; steering correctly; and full and proper observation.
In the early stages of your lessons on manoeuvring your instructor
will take care to select roads which are adequate and easy to cope
with. You cannot spend the rest of your life reversing and
manoeuvring only in easy places, however, so he will also arrange for
you to practise in one or two more awkward situations, perhaps in a
narrower road, or reversing downhill followed by uphill in the same

manoeuvre. Even worse, the first time you try it, can be reversing uphill and then downhill! The skill is in recognizing the change of camber and adjusting your vehicle control to suit.

Practical training: Department of Transport Syllabus Stages 6 and 7

There are four questions which every driver should ask himself before starting any manoeuvre which involves reversing or slow driving into a parked position. They are:

- Is it safe to do it here?
- Is it convenient to do it here?
- Is it legal to do it here?
- Is it within my capabilities?

If the answer to each of these is 'yes', then you may carry out the manoeuvre. But even so, and this especially applies to driving test practice, try not to spend too long in any one particular place practising. Apart from the boredom factor, you could also annoy local residents. One or two attempts at any one exercise in any one street should be ample.

As well as reversing to the left, you must be able to:

Reverse round a corner to the right, on level ground

Some instructors prefer to teach reversing to the right before reversing to the left. They argue, with merit, that it is easier to teach the needs and methods of observation by introducing the right reverse first. In either case, it is a question of teaching one method first, and then teaching the other by demonstrating the differences. The only points to bear in mind, however, are still the same three requirements:

- clutch control;
- correct steering:
- complete observation.

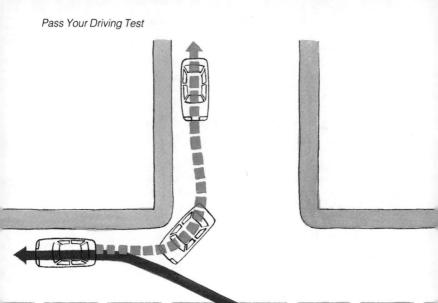

Reversing to the right. The blue line shows reverse gear

When reversing to the right it is essential to remember the need to look through the back window for the main part of the reverse in order to be aware of any other traffic on the road behind. Although it is possible to see the kerb better through the driver's window, this does present blind spot problems.

Having mastered reversing on the flat, you must be able to:

Reverse to the left and right, while on all types of gradient

Once this objective has been achieved, reversing should no longer present a problem. Perhaps the most complicated exercise you will do is one that involves reversing uphill to begin with, and then, as you begin to turn the corner, going downhill, with the potential for loss of clutch control. No doubt your instructor will find you a whole range of reversing exercises which will be much more difficult than any your Examiner will have available for his choice.

As well as reversing on different types of gradient, ranging from gentle slopes to steep hills, your instructor will also be able to find a whole variety of differing shapes and curves, once more ranging from gentle curves round a kerb to tight corners going at acute angles. Variety will also be shown in the width of the road available. At the end of this objective you will find it relatively easy to reverse in any situation, including your own driveway when the time comes.

Manoeuvring skills are now a large part of your driver training. In between them you will be driving through all kinds of traffic conditions, satisfied that you are able to cope with all roads without any other instruction from your supervisor, except for the directions of where you are to go. The manoeuvring objectives continue to show that you are able to:

> Turn the vehicle round in a confined space (such as a car park) correctly and safely, and under full control, correctly positioned, with maximum observation and care for all other road users

No longer is your instructor content to get you to perform your skills of manoeuvring in nice easy wide roads. Instead, he will encourage you to demonstrate your ability and confidence to park and manoeuvre in the sort of places which will be needed after the test is over and you are driving on your own. This will include parking-meter spaces, and you must therefore be able to:

> Park the vehicle using reverse gear, in any suitable confined space

From a driving test point of view this has been defined as reversing into a gap between two cars. Driving examiners, however, tend to use a parked car ahead of you, alongside which you should stop, then reverse into the gap behind that car. You will have already proved to your instructor your ability to park in reverse gear into a gap only about one and half times the size of your own car. This is your target.

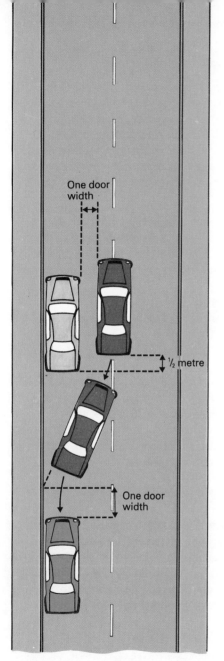

One door width

½ metre

One door width

Parallel parking in reverse gear

Similarly, you should be able to demonstrate that you are able to

Park between other vehicles

Park at a parking meter bay, both on the right and the left

This completes all of your objectives on manoeuvring. At this stage of your training, your instructor's role is simply one of giving you directions, and occasionally setting you a series of objectives, followed by a debriefing. Should any instruction or prompting be necessary, it may be that you should go back to some of the previous **objectives** and ensure that you can cope adequately without help.

Achievement of objectives

As you draw to the end of the third part of this training book, you will need to confirm (to yourself) that you have achieved the objectives laid down at the commencement of this part. You can either complete this check sheet yourself as a personal commitment that you have reached an acceptable standard or you can invite your instructor to confirm your ability to perform each of the following tasks to the standard acceptable to him or her.

At the end of this third part of the book I am now able to:

Recognize and cope with each of the following:
bend; corner; turning; junction; crossroad; and roundabout (including mini, normal, large and unusual);

Negotiate each of the above, safely and correctly: on approach, while on them and on leaving;

Recognize, identify and cope with all types of road and traffic signs, while driving in light traffic conditions;

Identify motorway signs and explain their meanings;

Show a high standard of knowledge of the Highway Code and other motoring matters;

Reverse in a straight line;

Reverse round a corner to the left, on level ground;

Turn the vehicle in the road, using forward and reverse gears, safely and correctly, and with due regard to all other road users;

Turn the vehicle in the road accurately, with regard to various widths of the road;

Reverse round a corner to the right, on level ground;

Reverse to the left and right, while on all types of gradients;

Turn the vehicle round in a confined space (such as a car park), correctly and safely, under full control, correctly positioned and with maximum observation and care for all other road users;

Park the vehicle, using reverse gear, in any suitable confined space;

Park between other vehicles;

Park at a parking meter bay, both on the right and on the left.

Signed: _____

Pupil: _____

Instructor: _____

Date: _____

Questions

Stage 3

1 A driver is approaching traffic lights showing 'amber' on its own. The driver will know that

(a) The next colour to show will be green

(b) The next colour to show will be red

(c) The lights will stay amber for at least ten seconds

(d) It is most likely that he will need to slow down or stop

(e) He should now use the mirrors and be prepared to signal

2 When approaching a left-hand bend in the road ahead a driver should normally position the vehicle

(a) As close to the left as possible

(b) In the centre of the lane presently occupied

(c) As far to the centre as possible

(d) Straddled across the centre line if safe to do so

(e) In the same position as if the road were straight

3 Seat belts are now fitted in the front and rear of all new motor cars. The laws about wearing them are quite specific.

(a) Children under the age of 14 must wear rear seat belts if they are fitted to the car.

(b) Drivers of motor cars are responsible and must ensure that their adult front seat passengers wear seat belts.

(c) Seat belts may be removed by drivers when reversing

(d) Driving examiners are required to wear seat belts when carrying out driving tests

(e) If a car's seat belts are dirty or inoperative a driving examiner may abandon a driving test.

Answers

1(a) All drivers must know the sequence of traffic lights. Green does not follow amber. Green follows red and amber together.

(b) Red follows amber, this is therefore correct.

(c) There is no way of knowing how long a colour will remain.

(d) It is indeed most likely that he will need to slow down or stop, so immediate action is required.

(e) Certainly the mirrors must be used, and a signal may well be needed, depending upon the action required, and the traffic conditions which exist.

2(a) When approaching a left-hand bend a driver should normally position his vehicle in the same place as for normal driving. Therefore he should not position it as close to the left as possible.

(b) The correct position should be in the centre of the lane presently occupied.

(c) It would be wrong, unnecessary and potentially dangerous to move to the centre line.

(d) Even if it appeared safe it would still be wrong and unnecessary to straddle the centre line. The reason for this is that the driver would be committed to following a line which is too wide. Should any other traffic appear a fresh line would need to be taken. This presents unnecessary problems, which could have been avoided by following the safety line.

(e) This is correct. You should follow the same line as if the road were straight, that is, following the normal safety line.

3(a) This is correct. The driver of any vehicle must ensure that all children under 14 are belted up in the rear seats of vehicles where seat belts are fitted.

(b) Actually adults (over the age of 14) are responsible for their own seat belts. Drivers are not responsible, but they can, and should, always refuse to allow them to remain in the car unless they do belt up.

(c) Seat belts can, and often should be, removed when a driver is carrying out any exercise which involves using reverse gear. They must be put back on immediately afterwards of course.

(d) Driving examiners are required to wear seat belts like all other passengers. However, like all other citizens, it is possible for them to be excused on medical grounds in some cases.

(e) If the seat belts do not work, or are too dirty to wear, a driving examiner is entitled to cancel or abandon a driving test.

Part 4
Lead-up to your test

Stages 8 and 9 require you to:

Understand and answer questions on night driving, adverse weather conditions, and to cope with them as they arise;

Cope with all kinds of urban and rural road traffic conditions;

Understand and answer questions on dual carriageways and motorway driving procedure;

Reassure the instructor that motorway driving will be coped with safely and correctly after the test is passed.

At the end of the tenth stage you will also be able to pass the Department of Transport Driving Test for Category B licence for motor vehicles, and specifically cope with the following requirements:

Cope with a simulated driving test route, with no serious or dangerous errors of any kind;

Show effective use of the mirrors;

Manoeuvre as required for the test;

Demonstrate making good progress;

Meet, cross the path of and overtake other traffic safely;

Maintain a correct safety line;

Act correctly at pedestrian crossings (all kinds);

Show awareness and anticipation of the actions of all road users;

Answer questions on any motoring matters, satisfactorily for the standard required of a new driver;

Identify any road or traffic sign and demonstrate the correct reaction;

Maintain full control of the vehicle at all times, and take charge of the traffic situation as required, making and taking all the required decisions in good time and correctly.

Introduction

Whereas the aim of driving instructors and examiners is that pupils should learn to drive as a life skill, the pupil's aim is generally much simpler: to pass the driving test. Indeed, 'passing the driving test first time' has an accolade of success attached to it far beyond its real value. For passing the test is easy, and passing the test first time is also quite achievable for every learner driver if only they pay attention to two basic facts.

These are, that the test is the simplest driving test you will ever take; and also that the Driving Examiner actually wants to pass you. When the test starts you have a clean sheet. You will only fail if during the half an hour or so driving you actually commit a serious or dangerous error. Provided all your training and practice beforehand have resulted in your being able to demonstrate your ability to drive competently then this should not be difficult. However, if your instructor is not yet confident, or if you have not been able to complete the progress chart, satisfactorily, then taking the test is a gamble. If you are lucky you will fail and the Examiner will point out where you went wrong; if you are not so lucky you might scrape through and make your next serious or dangerous mistake when you are on your own. That could be very unlucky, either for you or perhaps some other road user.

Some pupils who fail the test do so because they are not used to being with an examiner. They will tell you that had it been their instructor who was with them they would have been all right. This is rubbish of course. There are more than one million driving test failure excuses every year, and they all boil down to the same one: not yet competent to drive alone safely. What these drivers are actually saying is that they want to have an instructor sitting with them while they drive – with L-plates of course.

So the one question to ask yourself before you actually take the driving test is, 'Am I competent to drive alone, without instruction or guidance from anyone? Will I be safe under these conditions?' The answer to that is to show your instructor that you can do so without any further tuition, and certainly without the instructor needing to use, or even to contemplate using, the dual controls.

Ideally the driving test should be seen as a step along the way towards learning to become a competent and safe driver. If your instructor has planned your lessons according to a structured programme, you will find that the test is taken at about seven eighths of the way through the course. The final steps to be covered after the test would be regarded as important enough to require initial training, followed by constant practice. These final lessons, however, can be slotted in occasionally, so that it is possible to buy your car, use it for local journeys after the test, but still take sufficient professional training to ensure you are not taken unawares when you first venture out on to motorways, at night, or when the road and weather conditions are not so good.

The test itself should be applied for when both you and your instructor feel that you have achieved sufficient success in your training programme so that you can fit the rest of your training into the time which will be available before the test comes up. Driving test centres are grouped into sectors. Each sector has one centre at which vacancies for tests are available, if not on demand, then certainly at short notice. Your instructor will know which of the centres in your sector has a suitable waiting time.

Ideally a waiting period of about a month to eight weeks would be sufficient. Once you know your driving test date, your driving lessons take on a new sense of urgency. Quite often they also mark a deterioration of your abilities as a driver. Whereas before the test date was known you were quite happy to drive your instructor round any route he might select, as soon as the test date is known you are much more aware of your own mortality. Fears and doubts can bedevil you to such an extent that you make mistakes which would previously have never been made at all. This is quite usual, and you must simply ensure that you recognize what is the cause of the mistake and how you can come to terms with it. If, however, you find that you cannot get through this problem and the test is really worrying you, there is only one choice. Cancel the date you have and ask for another one perhaps four or five weeks later.

This is an appropriate occasion to use your personal check sheet in this book, for you need to confirm that all the objectives listed at the beginning of each section can really be achieved.

Your instructor will now be playing the role of the examiner. Each time you enter the car he will ignore you as a customer and simply watch how you behave as a motorist. He will check how efficiently you enter the car – whether you close the door safely and carry out the safety checks for entering and starting the engine – then he will tell you where he wants you to drive. If you need to be reminded what the safety checks are, you are not ready. At this stage, ask your instructor not to tell you what to do but only to prompt you as a last resort. What you don't want at this stage of your training is repeated instruction; you must now abandon the habit of waiting for your instructor to tell you what to do. The greatest single cause of unnecessary driving test failures is candidates automatically waiting for instructions from their examiners.

Instead of practising round all the various test routes which you will both know by now, you should be allowed to drive on every type of road in your area, including dual carriageways (but not motorways), thus gaining as much experience as possible. It is also helpful to be given a general itinerary, such as 'Drive from your home/driving school office to the Town Hall, enter a multi-storey car park, and await the next instruction.' This could be to leave the car park and make your way to the outskirts of town, following a signed route marked A62 to the next town.

You will discover two things: first that the tasks now being set are considerably more difficult to carry out than the mock tests you have been successfully performing; and second that these new exercises are just the things which you will be doing after you have passed your driving test and find that motoring is merely an ancillary part of your life.

Overtaking

Of the three situations which are the start of accidents we have said that overtaking is the most dangerous. You will always need to follow other traffic, and you have little control over the distance behind you that another vehicle will keep. It is also difficult to plan routes to avoid right turns. But overtaking is the one manoeuvre over which you have absolute control.

All drivers will sometimes have to draw out in order to pass other parked and stationary vehicles, but this is not, strictly speaking, overtaking. The act of overtaking is always taken with the other road user moving in the same direction as yourself. Therefore even if the other vehicle is travelling much slower than you wish you can still opt to drive at the same speed as he is doing. Overtaking is entirely up to you.

There are a number of occasions when it is illegal to overtake. You will find them listed in the Highway Code. Similarly you will also find a number of occasions when you may actually overtake on the left as well. Read and study all of these well. Not because you will necessarily be asked questions on them, but because you will have to put them into practice in the future.

There are also many times and places when it is not suitable to overtake. Your instructor will almost certainly spell these out for you. When you cannot see the road ahead is clear, or when the vehicle in front is too large or is driving at almost the same speed as yourself, are three obvious examples.

However, there will be occasions when you really will be able to overtake, safely and without inconveniencing any other road user, and not to do so would actually hold up other traffic. This is when you need to put your sequences into operation. The overtaking sequence is: P–S–L, M–S–M (i.e the reverse to the usual sequence).

This does not mean that you don't use your mirrors before you pull out. But having decided to overtake and having checked that it is clear behind you:

- get into a **position** where you can see ahead;
- make sure you have sufficient **speed** available.

- **look**, and assess the whole situation; but before you **decide**:
- use your **mirrors** to ensure it is still safe;
- Then **signal** your intention to move out (if necessary);
- and carry out the **manoeuvre**. This includes using your mirrors again especially before pulling back in to follow the safety line once more.

Overtaking is always a big step to take. Most learners spend a long time on their early lessons before they actually do it for the first time. Many spend much longer before they actually overtake on their own initiative. But you will have to learn how to do it, and how to do it safely every time.

If you are at all mathematically minded you will already appreciate that overtaking needs to take into account the combined speeds of oncoming traffic and the overtaking vehicle. What make even more interesting mathematical studies however are the total distances travelled at 60 m.p.h. by an overtaking vehicle when the vehicle being overtaken is travelling at 50 m.p.h., then 55 m.p.h., and finally 59 m.p.h.

You have to work out the distance from a safe spot, say 176 feet behind to a safe spot ahead, a further 176 feet, so it is the equivalent of driving 352 feet at 10 m.p.h., or 5 m.p.h. or 1 m.p.h. The distances required are considerable. Especially when you assume that any oncoming traffic, out of sight, will be only travelling at the maximum speed limit of 60 m.p.h. Sometimes they don't.

Ask your instructor to allow you to measure exactly how far you can travel when overtaking another vehicle travelling at only 5 m.p.h. less than you are.

Road and weather conditions and night driving

As well as the problems created by driving different types of vehicle, road and weather conditions also have a very strong effect on the way any vehicle handles. It is essential that your driver training covers a whole range of different types of road, including single and dual carriageways, and those with two, three or more lanes of traffic travelling in each direction.

Because most driving instructors seem to do some form of home collection, many clients benefit because a lot of their initial training takes place in the sort of areas where the learners will continue to drive after their tests are passed. Nevertheless, it is essential that instructors take their clients over as varied a range of roads and traffic conditions as they can. Nothing can be more futile than to have a client pass a test, without ever having been on any other roads except those used in test preparation.

This also highlights one of the few drawbacks of learning to drive through an intensive course where the client takes a five-, ten- or fifteen-day course of concentrated training. It can easily happen that throughout the whole of the training period (and possibly the test too) no rain or wet weather is encountered. So unless a very good driving instructor has been chosen, there is a possibility of a new driver taking to the roads who has never used the windscreen wipers, washers and demisting equipment. This can be disconcerting, but the real danger lies in the fact that they have never driven in wet conditions.

One of the greatest single dangers associated with driving in the rain is the effect of a sudden shower on a road which has been dry for a long time. During the long dry spell the rubber being worn away from the tyres mixes with the oil and greasy deposits from engines to create a gooey mixture which effectively fills in all the gaps of the tarmac road surface. Water is the natural lubricant for rubber, and when rain falls on to this surface it creates a very slippery surface for tyres. Aquaplaning and skidding are two problems which need to be borne in mind whenever such conditions exist. The best tyres in the world will not displace water quickly enough in such circumstances, and any sudden or harsh braking can have frightening effects. (See also pages 157, 167, 183 for more on weather conditions.)

As with all these topics, the object here is not only to inform the new learner driver of what care need be taken, but to remind him that other drivers and road users may also be taken by surprise. Road traffic incidents arise because drivers are either unaware of a particular danger, or consider they have more skill than they actually possess. Regrettably, most drivers assume they are able to stop their vehicles in much shorter distances than proves to be the case.

Research over recent years has proved that generally speaking male drivers overestimate their abilities to stop in any given distance, and female drivers have weaknesses in their spatial awareness.

These weaknesses are corroborated by men when driving at speed, and by ladies when they park their vehicles. This is not to say that no women drivers are good at parking, nor that all men tend to drive too closely in bad weather conditions. But statistically speaking these facts can be proven. One fact which can be seen to need no confirmation, however, is that where learner drivers – whether male or female – are taught proper car control, and given the opportunity to learn and practise situation control, they are less likely to be involved in accidents than those drivers who pick things up as they go along. There is no alternative to first-class training. Experience is a hard and expensive substitute.

If none of your lessons has taken place in the dark, you should also give serious thought to how you gain experience in this aspect of your driving. Ideally every learner driver should have at least two lessons in darkness. This means not only driving at night but also driving on roads where there is no street lighting, for urban street lighting is often almost as good as daylight. However, where the lighting is sparse there are often large pools of blackness which make driving more hazardous. Look out for pedestrians or cyclists who are not wearing anything white or reflective. Look at the road ahead, not at the headlights of oncoming cars, and remember that headlamps must be dipped if there is oncoming traffic, or if you are following another car (in both cases full beam could dazzle the other driver). Once more, the only safe rule for night driving, as with driving in daylight, is never drive faster than you can see.

Accidents and first aid

Two items which also need to be covered during this part of your training, hopefully in theory only, are what to do in the event of an accident, and basic first aid. It may be that your instructor is not the best person to teach you the latter (it's not what he's trained to do) but there are certain rules you can readily learn which may well save someone's life. Learning life-saving skills is not a compulsory part of driver training, but any effort to learn them is instantly rewarded should the occasion ever arise when they are put to use. And nothing can ever compensate for the feeling of total inadequacy should you ever be faced with such a situation and find yourself unable to cope.

None of this is to suggest that you should spend any of your driving lesson time tying bandages or taking walks alongside motorways! It is important, however, that you appreciate at this stage the need to obey the law in all its demands upon you as a road user, and that you recognize your responsibilities to your fellow road users at all times. It is for this reason that you should also recognize the penalties which can be incurred, both those legally imposed, and those which may result from accidents. The motoring courts are not set up to extract money and give penalty points to guilty motorists; they are used to act as a deterrent to those who may wish to use their vehicles and the roads selfishly and without thought to others who may be there. (For further advice on accidents etc, see pages 59–61 and 188.)

Warnings

It is just as important to know what to look for in the way of warnings of impending danger, both in the car and on the road. Very few incidents ever happen suddenly. There are plenty of warning signs which can be seen or heard in advance for those who are prepared to look and listen. The warning lights and gauges on your dashboard are an example. Before your radiator boils over the temperature gauge will have risen dramatically. Before you run out of fuel the gauge will have given you warning. Sometimes you will be given added warning by the use of flashing lights to reinforce the message. Tyres often become heavy to steer before they become completely flat. As with all warnings, the secret is to look for, and be aware of, any changes in the way the car behaves.

One final warning at this stage of your tuition. Everything you are learning is based on the premise that you intend to become a good, safe, competent driver. Unfortunately, not all of your fellow road users feel the same way. Therefore one of your earliest lessons is to accept fools and their actions patiently. The only competitive element in your driving should be yourself against the perfect drive you are trying to demonstrate – *never* one of speed or space between you and any other road user. If someone else decides to overtake and cut in, or turn in front of you without a signal, there can be no excuse, morally, in front of a judge, or to a coroner, for taking revenge, or trying to punish the offender. If you ever feel the need for a streak of competition in your driving, make it that of seeing how long you can drive without making a single mistake.

Objectives: Department of Transport Syllabus Stages 8 and 9

You should now be, to all intents and purposes, a competent driver who can manoeuvre the car safely and accurately, with complete control, and cope with any road and traffic conditions you may encounter. In these final stages, leading up to your driving test, you need to practise these skills as often as you can, and ensure that you have a good understanding of the Highway Code and all matters and conditions relating to driving.

You must be able to:

> Understand and answer questions on night driving, adverse weather conditions, and cope with them as they arise

Whether or not you have practical experience of these conditions, make sure you know, at least in theory, how to act. (See pages 120, 128, 169–70 and 183 which deal with these conditions.)

You must be able to:

> Cope with all kinds of urban and rural traffic conditions

Read carefully those sections in Part Five that deal with this topic, though you will find that, in the course of your driving instruction, you will have learned correct situation procedure.

You must also be able to:

> Reassure the instructor that motorway driving will be coped with safely and correctly after the driving test has been passed

Although you cannot, of course, yet drive yourself on any motorway, read with attention the section on motorways (pages 160–66) in Part Five, and make sure that you recognize motorway signs and understand how such systems work.

Ideally, all learner drivers should book a post-test course of practical driver training to include motorway, hazardous weather, night-time driving and long-distance driving, though in practice this can be difficult to put into effect. So unless you are sure that you will be able to invest the time, energy and minimal expense in this extra tuition, it is essential that your instructor is satisfied that you will cope when you are left to your own devices.

The driving test: essential preparation: (Department of Transport Syllabus Stage 10)

The tenth stage of learning to drive is perhaps the simplest, and yet it is one that sometimes worries learners more than any other: it is to prove your knowledge and skill to another person – a Department of Transport Driving Examiner. Yet the Examiner will not expect anything of you that you have not already been taught, and no one will ever be as critical of your driving as your own instructor!

Passing the driving test should be a simple question of putting into practice all that you have been taught. There is nothing in the driving test that cannot be coped with easily by someone who has been correctly taught and prepared, and by someone who has proved that they can achieve each of the driving objectives presented so far. Nevertheless, it is very useful to spend the final four or five lessons, perhaps in the two weeks or so leading to the test, in test preparation. During these lessons the Instructor will take over the role of the Driving Examiner, even to the extent of calling your name formally and using only the vocabulary of the Driving Examiner.

You must be able to:

Cope with a simulated driving test route, with no serious or dangerous errors of any kind;

Show effective use of the mirrors;

Manoeuvre as required for the test;

Demonstrate making good progress;

Meet, cross the path of and overtake other traffic safely;

Maintain a correct safety line;

Act correctly at pedestrian crossings (all kinds);

Show awareness and anticipation of the actions of all road users;

Answer questions on any motoring matters, satisfactorily for the standard required of a new driver;

Identify any road or traffic sign and demonstrate the correct reaction;

Maintain full control of the vehicle at all times, and to take charge of the traffic situation as required, making and taking all the required decisions in good time and correctly.

If you can do all of this correctly to the satisfaction of your driving instructor there is nothing that the Driving Examiner can ask you to do that will catch you unawares. If, during your build-up the driving test, you find that certain aspects are letting you down, you will need to go back to the specific objectives which are weak, and ensure that you are now capable of achieving each of them effectively.

Achievement of objectives

As you draw to the end of the fourth part of this training book, you will need to confirm (to yourself) that you have achieved the objectives laid down at the commencement of this part. You can either complete this check sheet yourself as a personal commitment that you have reached an acceptable standard or you can invite your instructor to confirm your ability to perform each of the following tasks to the standard acceptable to him or her.

At the end of this fourth part of the book I am now able to:

Reverse to the left and right, while on all types of gradient;

Turn the vehicle round in a confined space (such as a car park) correctly and safely, and under full control, correctly positioned, with maximum observation and care for all other road users;

Park the vehicle using reverse gear, in any suitable confined space;

Park between other vehicles;

Park at a parking meter bay, both on the right and the left;

Cope with all kinds of urban and rural traffic conditions;

Understand and answer questions on dual carriageways and motorway driving procedures;

Reassure the instructor that motorway driving will be coped with safely and correctly after the driving test has been passed;

Cope with a simulated driving test route, with no serious or dangerous errors of any kind;

Show effective use of the mirrors;

Manoeuvre as required for the test;

Demonstrate making good progress;

Meet, cross the path of and overtake other traffic safely;

Maintain a correct safety line;

Act correctly at pedestrian crossings (all kinds);

Show awareness and anticipation of the actions of all road users;

Answer questions on any motoring matters, satisfactorily for the standard required of a new driver;

Identify any road or traffic sign and demonstrate the correct reaction;

Maintain full control of the vehicle at all times, and to take charge of the traffic situation as required, making and taking all the required decisions in good time and correctly.

Signed:

Pupil: _____

Instructor: _____

Date: _____

Questions

Part 4

1 A driver applies the brakes of his vehicle at the same time as steering round a bend to the right. It is true to say that

(a) The weight would be thrown to the front of the car

(b) The weight would be evenly distributed on all four tyres

(c) The front nearside wheel would take most of the weight

(d) A danger of skidding would arise

(e) The car would be in its most stable position at this point

2 During the Department of Transport Driving Test, which of the following manoeuvres can a candidate expect to be asked to do?

(a) A genuine emergency stop

(b) A reverse to the left

(c) A reverse to the right

(d) A turn in the road

(e) A down-hill start

3 When you are driving in bad weather conditions in the winter you must remember

(a) That braking distances may take longer

(b) To use your headlights on main beam during fog

(c) To stick to main roads where possible in snow

(d) Not to use your windscreen washers when it is freezing

(e) To switch off front and rear fog lights when they are not needed.

Answers

1(a) When braking at any time the weight of the car is always thrown to the front of the vehicle.

(b) The weight is only evenly distributed when the car is being neither accelerated nor braked. Any change of speed or direction will alter the weight distribution on the tyres.

(c) In this instance the front nearside (that is the passenger side) wheel would take the most weight of the car.

(d) A danger of skidding will always arise when a vehicle is braked and the front wheels are turned. Always try to avoid steering and braking at the same time.

(e) The car is in its most *un*stable condition when braking and steering take place together. This is why it is so important to complete your braking before you start to turn your wheels.

2(a) During the driving test you would not normally expect to do a genuine emergency stop, although you must always anticipate an emergency situation arising. Nevertheless you would not be *asked* to do a genuine emergency stop.

(b) A reverse into a limited opening to the left is one of the standard exercises that all learner drivers are asked to do on test.

(c) A reverse into a limited opening to the right is not normally asked; however if the vehicle has no side windows, or if the Driving Examiner is unable to find a suitable road into which to reverse to the left, he is entitled to ask a candidate to reverse to the right instead. You must be prepared for this exercise. Apart from any other consideration, after you have passed your test you may find you prefer this method of changing direction because it can be safer.

(d) A turn in the road is a standard driving test requirement. Incidentally in many European countries this exercise is not allowed. The turn in the road exercise is really practice for when you need to park your vehicle in any confined space. It is not normally used as a means of changing direction.

(e) Assuming you live in an area of the country where there are hills and cambers of any sort, you must expect to be asked to move off on the level, uphill and also downhill. Remember that when moving off downhill the sequence is different. You are expected to use your footbrake instead of the hand or parking brake.

3(a) The braking distances in wet weather are normally doubled; in icy conditions they can be ten times as far. Even such things as wet leaves and bad road surfaces generally increase braking distances considerably. If the weather is bad allow a lot more space not only between you and the vehicles ahead, but for unthinking drivers behind you too.

(b) You shouldn't normally use your main beam headlights in fog, because the water in the air tends to reflect the light straight back at you. Use dipped lights however, not only to help you see further, but also to ensure that you are seen.

(c) Staying on main roads during snow makes sense. Try to put your front tyres into the same tracks that other vehicles have made, rather than create fresh routes for yourself. If you have to drive on fresh snow, try to avoid stopping and starting too much. The secret is to move slowly, without picking up or losing too much speed.

(d) Using your windscreen washers when it is freezing can result in a sheet of ice or slush on your screen, making it impossible to see through. Clean your screen as often as you can. If it is not actually snowing or raining try to keep the screen dry. Use de-icing fluid mixed with your screenwasher liquid if you can; but even this can freeze when the weather is really cold. If necessary pull off the road and clean your screen with a scraper.

(e) Rear fog lights should only be used in extremely bad driving conditions when they help following traffic to see you. Only use yours if the visibility ahead is so poor that you are grateful to see other traffic using theirs too. If ever you see anyone else's rear fog lights on and it is unnecessary, check to see that you have not left yours on too.

Part 5 The driving test: Pass first time

Perhaps the most common question to follow any introduction of the subject of driving tests is, 'Did you pass first time?' Never 'How much did it cost you?' nor 'How long did it take?' Always 'Did you pass first time?' Professional driving instructors deplore this attitude, since it encourages learners to rush things, or to take a test 'for the experience' with the likelihood of failure. This doesn't help the instructor's record nor his standing with the examiners, and it doesn't do much for the learner's confidence. Any good instructor can sit in a car with any learner driver and take him for a mock test, at the end of which he can give as accurate an assessment of the learner's driving as can any Driving Examiner, though he cannot, of course, issue a pass certificate. And if the learner fails, at least he hasn't had to pay over £20 for the privilege of being told he is not up to standard and will have to wait another month at least before he can try again.

The purpose of the driving test is to see if you are safe to drive your vehicle on the roads on your own. If the Examiner thinks you *are* safe, you pass; if you commit a serious or potentially dangerous fault then you will have to retake the test. If you can satisfy your instructor on a couple of mock tests then you should also be able to satisfy the Examiner, for your instructor will almost certainly have taken you round a few routes which are considerably worse than those the Driving Examiner is allowed to choose.

As I have said before, the only danger is that you may have become too reliant upon your instructor, perhaps waiting for him to remind you to use your mirrors before making a signal. So be advised, in your final lessons, to ask your instructor to role-play the Examiner for thirty minutes or more at a time. That way, you'll be used to the way examiners speak, the way they seem to ignore you, the way they sit and write little comments on a pad on their laps.

Let us now simulate a driving test. What will the Examiner ask you to do?

The test lasts about half an hour. The Examiner will first meet you in the waiting-room, ask you to identify yourself and request that you lead him out to your car, where he will test your eyesight. If you are taking the test in a normal motor car you will need to read a conventional (7-figure/letter) number-plate at 20.5 metres (67 feet). Once you are in the vehicle the examiner will ask you to drive round a pre-set route during which time he will direct you as you go along and ask you to carry out a number of manoeuvring exercises.

The test begins with a short drive and then the Examiner will ask you to pull in at the kerbside at a convenient place. This start to the test, called the natural drive, is intended to help you to relax a little, and also to get you into an area away from yellow lines. This will help you find a spot which is convenient for pulling in whilst the emergency stop exercise is explained.

The test itself is standardized throughout the country. In each town a number of test routes are laid down and each Examiner determines in advance which route he will take. The marking sheet used by the Examiner is laid out in a simple form so that all he has to do is assess each move you make and compare it with a standard that has been laid down as acceptable. Each mistake that you make is assessed at the time you make it and classified into one of three categories: Minor, Serious or Dangerous. Only serious or dangerous errors actually result in failure. A serious or dangerous error, by definition, has to be one which has caused another vehicle or road user to be inconvenienced or put into danger.

At the time of marking the fault, the Examiner is required to ask of himself 'If this were to be the only fault made by the candidate throughout the whole test would it justify failing on this one item?' If the answer is yes, then it is marked X for serious or D for dangerous. If not, then it is either marked / as a minor error, or disregarded. (See pages 134–47 for examples of these marking sheets, the specified wording to be used by all examiners, and other related material.)

At the end of the test the Examiner will look at his marking sheet and if there are no X or D markings you will pass. If you do fail, however, he will explain to you exactly what you did wrong and then fill in the failure sheet for you.

The Highway Code questions are asked at the end of the practical test. Generally speaking an Examiner asks five or six questions from a list he has prepared. (See pages 138–39 for examples.)

The specific items the Examiner will test are:

Checks before starting If you carry out the safety checks when you enter, and each time you start or re-start the engine, the Examiner will be satisfied.

Using the controls The Examiner will check to see that you are using the accelerator, brake pedal and clutch correctly and smoothly. He is especially concerned to see the way you balance the use of the accelerator with the clutch when you move off. Coasting (by driving with the clutch down) is potentially dangerous, and would be marked so.

With regard to the hand controls he notes especially that you select the gears correctly and in good time, without taking your eyes off the road. He looks to see that you steer in a safe, controlled way, following the safety line. He also wants to see you use the handbrake, or parking brake, correctly whenever you are stationary. If you have time to ask yourself if it is needed, it is.

Moving away He will see if you can move off safely from a variety of places, including from behind another vehicle, and on both up- and down-hill slopes. He is looking to see if you use your mirrors and check over your shoulder for blind spots before you decide to move.

The emergency stop This is usually tested early, in case candidates are worried about doing it. However, you will have had a lot of practice and know exactly what to do. React quickly, and brake to a controlled stop. Avoid locking the wheels and steer straight. Only put the clutch down at the last moment. If you stall he won't mind, and it gives you the chance to demonstrate your starting procedures again. Remember that in wet weather it is easier to lose your grip. If necessary release the footbrake for a moment and brake again to a stop.

Reversing You will be quite proficient at this. Make sure you keep the car under full control, that you steer correctly as you turn the corner, and above all that you look where you are going most of the time, with occasional glances all round to make sure it is safe and you are not inconveniencing any other road user.

Turning in the road As with the reversing exercise you will have had plenty of practice, and will remember the three points that need to be taken into account: clutch control; effective steering; and full observation and care for other road users.

Parking in reverse gear You will have practised parking in reverse gear both to the left and the right. However, the Examiner will only ask you to demonstrate your ability to park behind a single car (with plenty of room behind you). Aim to park close to the kerb and within two car lengths. As with the other manoeuvres, the secret is to look carefully for other road users, especially traffic which could be held up by your front swinging out.

Using the mirrors Use your mirrors often, so that you have full knowledge at all times of what is around you. If necessary look around you as well. But looking is not enough; act sensibly on what you see in your mirrors. Don't forget that 'door' mirrors are sometimes called 'life-savers'; they enable you to predict what might otherwise prove an embarrassing moment. Use your mirrors well before you signal, change direction, turn, overtake or change lane, slow down or stop, or open any door. M–S–M is your standard routine by now of course.

Giving signals Use the correct signals, and only those. Indicators are the best way to signal, but remember that brake lights and arm signals can be used to good effect too. M–S–M is still the correct sequence. If you do use indicators, be especially careful that you use the correct one, and remember to cancel it after use.

Action on signs and signals Obey all traffic signs and also road markings. Keep to the left whenever you can, and always stay in the safest and correct lane. Nevertheless watch out for and obey markings which tell you to follow a specific lane. Obey all traffic lights and all traffic controllers, and also act sensibly on what you see other road users doing.

Care in the use of speed Don't drive too fast for the road conditions, or break the speed limit. Use of speed also refers to the safety distance you maintain behind other traffic. On wet roads this distance needs to be greater than for dry ones.

Making progress This is where many learner drivers go wrong when they freeze up. Keep up with traffic flow when it is safe to do so and try not to be hesitant at junctions. Proceed when it is safe, and do not hold up other traffic unnecessarily.

Road junctions Adjust your speed on approach correctly. The sequences of M–S–M and P–S–L both apply. If you approach at the correct speed you stand a better chance of being able to make the correct decision about stopping or going. Maintain the safety line for left or right, and watch out for other road users who may encroach upon your piece of road. Take effective observation on approach and before emerging.

Overtaking, meeting and crossing other vehicles Allow at least a door width when overtaking moving or stationary traffic. Allow at least the same room for oncoming traffic as well. Allow more room for cyclists if they appear to be wobbling at all; otherwise allow as much room as for a car. If the road is restricted in width, drive slower; if necessary slow right down or wait until the other traffic has gone. When you turn right in front of any oncoming traffic make sure you have plenty of time and room to clear. If in doubt, wait.

Normal position on the road Keep well to the left in normal driving, maintaining a safety line. If lanes are clearly marked, stay in the centre of them.

Passing stationary vehicles Look out for anyone who might step out between parked vehicles, especially young children or pushchairs; or doors opening.

Pedestrian crossings These include gaps between studs at traffic lights as well as pelican and zebra crossings. If a person might step on to a crossing, you must expect them to. If the crossing is not a controlled one, encourage them to cross by use of intelligent slowing down arm signals. Do not wave them over, though.

Select safe position for normal stops When the Examiner asks you to select a convenient place to pull in, choose the most suitable one you can see. Avoid inconveniencing other road users. However, do not just keep driving on because you don't see the perfect spot; select the best available.

Awareness and anticipation These are the items marked by examiners when none of the previous items appear to cover what you have done incorrectly or unsafely. Anticipation is the sign that you are planning your driving by looking ahead. Try to be aware of everyone else around you and watch out especially for pedestrians or cyclists who may not be obeying the normal traffic rules. Try not to be taken by surprise. If necessary be prepared to take the initiative if you can.

The thirty or so minutes of the test will allow the Examiner to take you round the selected route. Listen carefully to where you are told to go as his directions will always be precise. Remember that he will want to get back safely too. The questions at the end are to enable him to assess your theoretical knowledge, and to confirm one or two things about the drive. He will not tell you if your answers are correct or not, but simply move on to the next question. But if you were safe, and competent – you will pass, *First time!*

Driving examiners' test wording

Driving Examiners are required to stick to a very precise script when they carry out driving tests. The reasons for this are to enable the Driving Standards Agency to ensure that all driving tests are carried out exactly the same, and that everyone stands the same chance of passing no matter at which test centre, or with which Driving Examiner, they are tested. This script should not be varied, although with experience individual words may be changed, provided that the sense of the script remains the same. However, this can cause some driving-test candidates a small problem if they are not used to hearing the words used as part of their training. It is for this reason that all driving instructors are advised to use the same vocabulary as that taught at the Driving Examiner Training Establishment. And why it is repeated here.

Practical part of test

Preliminaries

Good morning/afternoon Mr/Mrs/Miss . . .

Would you sign against your name, please.

May I see your driving licence, or some other proof of identity, please?

Will you lead the way to your vehicle, please.

Have you any physical disability that isn't mentioned on your application form? (Only if a possible disability is noted.)

Which is your vehicle, please?

Will you get into your car now, please.

General directions

Follow the road ahead unless the traffic signs direct you otherwise, or unless I ask you to turn, which I'll do in good time.
Move off when you're ready, please.

Would you pull up on the left at a convenient place, please.

Pull up along here, just before . . . please.

Drive on when you're ready, please.

Take the next road on the right/left please.

Will you take the second road on the right/left please.
(If necessary they will add: This is the first.)
At the end of the road turn right/left please.
At the roundabout:
take the next road off to the left, please.
take the road leading off to the right, please.
follow the road ahead, please.
(Further information may be given to assist your route through a hazard, such as: It is the second/third exit etc.)

Emergency stop

Pull up on the left at a convenient place, please.
Very shortly I shall ask you to stop as in an emergency; the signal
will be like this. (He will demonstrate by slapping the windscreen
with his note pad or sheet of paper.) When I do that, stop
immediately and under full control . . . as though a child had run off
the pavement.
Thank you. I won't ask you to do that exercise again.

Body of the test

LEFT HAND REVERSE
Pull up along here just before you reach the next road on the left,
please. I should like you to reverse into this road on the left. Drive
past it and stop. Then back in and continue to drive in reverse gear
for some distance. Keep reasonably close to the kerb.

RIGHT HAND REVERSE
Pull up on the left before you reach the next road on the right,
please. I should like you to reverse into that road on the right.
Continue driving on the left until you are past it. Move across to the
right and stop. Then back in and continue to drive in reverse gear
well down the side road, keeping reasonably close to the right hand
kerb.

PARKING AT THE KERBSIDE IN REVERSE GEAR
I want you to move off, stop alongside that car in front, and then
reverse back into this gap, taking no more than two car lengths to do
so, finishing up close to and parallel with the kerb. Move off when
you are ready.

*Note that since 1 April 1991, all learner drivers will have been faced with
the task of demonstrating their parking skills using reverse gear before they
pass their driving tests. This extra requirement has been brought in to
conform with one of the items listed in the EC's Directive on Harmonized
Driving Tests.*

*The parallel parking exercise itself will be one of three 'reversing
manoeuvres' which can be tested. Normally the Examiner will ask the
candidate to perform only two out of the three. (The others are reversing
round a corner and the turn in the road – often referred to as the 3-point
turn).*

The Examiner will conduct the reverse parking exercise by getting the candidate to stop at the kerbside two or more car lengths behind another parked vehicle.

The candidate will be marked on observation, manoeuvring skills and finishing position. All good driving instructors will have been teaching this exercise for many years; however now that it is a test requirement all learners should ensure that they get plenty of good professional training in how to carry out the exercise, and also practise their skills whenever possible. Not only will this give them more confidence when they are driving, it will help keep traffic flowing, and make maximum use of available parking space.

TURN IN THE ROAD

Would you pull up on the left just past the . . . please.
I'd like you to turn your car round to face the opposite way, using your forward and reverse gears. Try not to touch the kerb when you're turning.

ANGLE START

Pull up on the left just before you get to the stationary vehicle please. (If necessary he will add: Leave enough room to move away.)

At the end of the practical drive the Examiner will instruct you to pull in at whatever parking space is available, and then ask you the following:

Now I should like to put a few questions on the Highway Code and other motoring matters.

That's the end of the test and I'm pleased to tell you that you've passed.

or

That's the end of the test – I'm sorry you haven't passed, but your driving hasn't reached the required standard. If you'll give me a few moments I'll try to explain where you went wrong, and to help you I'll also mark the points to which you should give special attention.

Theoretical part of the test

For more than fifty-five years the oral part of the driving test has been conducted in the same way. At the end of the practical part of the test, the Driving Examiner asks you to switch off the engine.

He will then ask you five questions on the Highway Code and Other Motoring Matters, and he will show you six pictures of road and traffic signs from the booklet in his briefcase.

The questions can be based on almost anything with a motoring connotation, but Examiners are strictly controlled in the way that the question is put, and are not allowed to give any indication as to whether the answer given is correct or not. They are told simply to respond with the words, 'thank you' or 'I see'.

The following is a list of the areas around which questions may be asked and some examples of typical questions:

(a) Driving along

1 For what vehicles should you keep a special look out when overtaking or turning?

2 How should you use your headlamps when meeting or following other vehicles at night?

3 Under what circumstances are you advised to give way to buses?

(b) Safety of pedestrians and animals

4 What should you do when a school crossing patrol displays a 'Stop Children' sign?

5 What is the meaning of the zig-zag markings at pedestrian crossings?

6 Why are young and elderly pedestrians particularly at risk?

(c) Overtaking and lane discipline

7 Name some of the circumstances when you shouldn't overtake.

8 How does a bus lane operate?

9 What should you not do in traffic hold-ups?

(d) Mirrors and signals

10 What advice does the Highway Code give about signalling?

11 What precautions should you take before you move off, reverse, or open your door?

12 What should you always remember to do after using indicators?

(e) Road junctions (including roundabouts)

13 How should you approach a road junction?

14 What advice are you given about crossing or turning right into a dual carriageway?

15 What is the meaning of the criss-cross yellow lines marked at a road junction?

(f) Stopping, parking and reserving

16 What general precautions should you take before reversing?

17 When may you switch on your hazard warning lights?

18 What does the Highway Code say about reversing from a side turning?

(g) Motorway driving

19 How does a motorway differ from other roads?

20 How should you normally join a motorway?

21 When should you use the right-hand lane of a 3-lane motorway?

22 What should you do if you have a breakdown on a motorway?

23 If you see flashing red signals over your lane, what should you do?

(h) Level crossings

24 What does the Highway Code advise you not to do at any level crossing?

25 At a level crossing with no gates, barriers or attendant you may see flashing red lights. What do they mean?

The traffic signs chosen by the Driving Examiner for the candidate to identify are taken from the Highway Code.

Questions are occasionally asked about things which may have arisen during the drive, although normally no Examiner will ask questions on an item which is likely to be marked on a failure sheet. Questions are also occasionally asked about driving at night, in bad weather conditions, or about ownership of a motor car.

DL 25 (Rev 01/91)

Driving Test Report

Centre Anytown

Date 3 / 2 / 92

"Safe driving for life"

Candidate's full name Mary J Nameless

Name of School (where known) First Pass S.O.M.

Particulars of vehicle P C ✗

Category B

		Make	Peugeot
		Type	309
		Reg Mark	H243 OPO
		Year	19 91
		Time	

FET		1

HC		2

PRE		3

ACC	X	④
CL	X	④
G		4
F.BR		4
H.BR	X	④
ST		4

MO	PRE		5
	CON		5

ES	╱		6
	FR.BR		6

RV	CON		7
	OBS		7

TR	CON		8
	OBS		8

PK	CON		9
	OBS		9

MIR	SIG		10
	DIR	X	⑩
	ST		10
ROB	SIG		10
	DIR		10
	ST		10

SIG	O		11
	W		11
	L		11

SNS	ST		12
	DIR		12
	NE		12
	RM		12
TRA	L	X	⑫
	CON		12
SIG	ORU		12

PRO	+		13

PRO	−		14

J	SP+		15
	OBS		15
	POSR		15
	POSL		15
	RCC		15

OT		16
MAT		16
CAT		16

POSN		17

SH.V		18

PX		19

NS		20

AA	PED		21
	CYC		21
	DRI		21

ETA	V	
	P	

An example of a driving test report

Weather Conditions

Dry and sunny after light shower

Route Number 4

Brief description of candidate

Slightly built lady, mid-30s, wearing blue jeans, white blouse and with short brown hair.

Remarks

④ x Candidate unable to coordinate the accelerator with the clutch and handbrake whilst trying to carry out hill start exercise.

④ x Candidate drove with the handbrake still applied and warning light showing for considerable distance.

⑨ x Candidate failed to check mirrors before moving out to take a right turn position. Motorcyclist forced to brake hard.

⑪ x High street. Continued accelerating towards traffic lights which changed from green to amber - crossed stop line well into amber phase.

D 10 No

DL 24 No 1234567

Examiner's Signature

A. Tester

Disability Tests - Including eyesight failures

Driver Number

Description of any adaption fitted

Explanation of abbreviations on form DL25

FET	1	Unable to meet the requirements of the eyesight test.
HC	2	Unsatisfactory answers to questions on the Highway Code.
PRE	3	Failure to take proper precautions before starting the engine.
ACC	4	Uncontrolled or harsh use of the accelerator.
CL	4	Uncontrolled use of the clutch.
G	4	Failure to engage the gear appropriate to the road and traffic conditions or for junctions. Coasting in neutral or with the clutch pedal depressed.
F BR	4	Late and/or harsh use of footbrake.
H BR	4	Not applying or releasing the handbrake when necessary.
ST	4	Erratic steering, overshooting the correct turning point when turning right or left, both hands off the steering wheel, hitting the kerb when turning left. (NOTE control faults should *not* be marked '4' if committed at PK TR, REV, MO, or ES.)
MO PRE	5	Failure to take effective precautions before moving away.
MO CON	5	Inability to move off smoothly: straight ahead, at an angle, on a gradient. (Not to be duplicated at item 4.)
ES	6	Emergency stop; inadequate braking, slow reaction or lack of control. (Not to be duplicated at item 4.)
RV CON	7	Incorrect use of controls when reversing. (Not to be duplicated at item 4.)
RV OBS	7	Lack of effective observation during the reversing exercise.
TR CON	8	Incorrect use of the controls when turning in the road. (Not to be duplicated at item 4.)
TR OBS	8	Lack of effective observation during the turn in the road exercise.
PK CON	9	Incorrect use of controls when reversing.
PK OBS	9	Lack of effective observation during the parking exercise.
MIR SIG	10	Failure to make effective use of the mirrors before signalling.
MIR DIR	10	Failure to make effective use of the mirrors before changing direction.
MIR ST	10	Failure to make effective use of the mirrors before slowing down or stopping
SIG O	11	Omitting a necessary signal.
SIG W	11	Signal not in accordance with the Highway Code. Failure to cancel direction indicator. Beckoning pedestrian to cross.

SIG L   11   Signal too late to serve any useful purpose.

SNS ST   12   Failure to comply with 'STOP' signs, including 'STOP Children' sign carried by school crossing patrol.

SNS DIR   12   Failure to comply with directional signs.

SNS NE   12   Failure to comply with 'NO ENTRY' signs.

SNS RM   12   Failure to comply with road markings, eg double white lines, box junctions.

TRA L   12   Failure to comply with traffic lights (not pelican crossings).

TRA CON   12   Failure to comply with signals given by a police officer, traffic warden, or other persons authorized to direct traffic.

SIG ORU   12   Failure to take appropriate action on signals given by other road users.

PRO +   13   Driving too fast for the prevailing road and traffic conditions.

PRO −   14   Unduly hesitant or driving too slowly for the prevailing road and traffic conditions.

J SP +   15   Approaching junctions too fast.

J OBS   15   Not taking effective observation before emerging at junctions.

J POS R   15   Incorrect positioning before turning right.

J POS L   15   Positioning too far from kerb before turning left.

J RCC   15   Cutting right hand corners.

OT   16   Overtaking or attempting to overtake other vehicles unsafely.

MAT   16   Not showing due regard for approaching traffic.

CAT   16   Turning right across the front of traffic closely approaching from the opposite direction.

POS N   17   Incorrect positioning of the vehicle during normal driving (normally only a fault when other drivers are inconvenienced).

SH V   18   Passing too close to stationary vehicles.

PX   19   Failure to give precedence to pedestrians on a pedestrian crossing. Non-compliance with traffic lights at a pelican crossing.

NS   20   Normal stop not made in a safe position.

AA PED   21   Not anticipating the action of pedestrians. (Including inconveniencing pedestrians actually crossing the road at a junction whether or not controlled by traffic lights.)

AA CYC   21   Not anticipating the action of cyclists.

AA DRI   21   Not anticipating the action of drivers.

ETA V   Examiner took action by interfering verbally with the candidate's driving in the interests of public safety.

ETA P   Examiner took action by interfering physically with the candidate's driving in the interests of public safety.

Faults – definition and method of marking

A minor fault is one which does not involve a serious or dangerous situation.

1 When a minor fault is committed, draw a short diagonal stroke to the right of the appropriate abbreviation.

2 When a minor fault occurs a second time draw a further diagonal stroke to the right of the first. No *further* stroke must be added, irrespective of the number of times the fault is committed.

ACC	/	4
CL		4
G		4
F.BR	//	4
H.BR		4
ST		4

3 A fault, properly assessed as minor, however often repeated, must not be 'built up' into a serious fault warranting the marking 'X' and will not entail failure.

A serious fault is one which involves *potential* danger.

4 When a serious fault is committed mark with a cross.

Do not convert any previous marking to a cross but add a cross alongside.

RV	CON		7
	OBS	✕	7

MO	PRE		5
	CON	/ ✕	5

5 A fault assessed as serious and marked 'X' entails failure.

A dangerous fault is one which involves *actual* danger.

6 When a dangerous fault is committed mark with a capital 'D'.

Do not convert any previous marking but add a capital 'D' alongside.

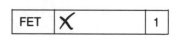

7 A fault assessed as dangerous and marked 'D' entails failure.

8 Where a candidate fails in the eyesight test the Examiner should put a cross to the right of 'FET'.

FET	✕	1

9 Where the candidate fails to satisfy the Examiner that he is fully conversant with the contents of the Highway Code a cross should be placed to the right of 'HC'.

HC	✗		2

10 Where the Examiner has to take action in a dangerous situation he should put a cross to the right of 'ETA V' or 'ETA P'. The nature of the action taken (whether verbal or physical) should be outlined briefly under 'REMARKS'.

ETA	V	✗
	P	✗

This does not apply to verbal directions to get the car moving.

New questions for 1994

By July 1994 learner drivers will probably have to answer questions in the following multiple choice format. This is typical of what will be asked and only one answer is correct.

Your are braking gently. Suddenly your car starts to skid, its rear going to the left.
What is the *first* thing you should do?

(a) Steer to the left ☐

(b) Steer to the right ☐

(c) Press the brake pedal firmly ☐

(d) Press the accelerator pedal firmly ☐

Which one of the following is a correct national speed limit?

(a) 70 mph on motorways for a car towing a caravan or trailer ☐

(b) 60 mph on dual carriageways for heavy goods vehicles above 7.5 tonnes ☐

(c) 60 mph on single carriageways for cars and motorcycles ☐

(d) 50 mph on single carriageways for buses and coaches ☐

In which one of the following is the policeman giving the signal to come on?

A ☐   B ☐   C ☐   D ☐

Look at the picture. This is what you see from your driver's seat.

What should you do now?

(a) Toot your horn to warn the riders of your approach ☐

(b) Slow down and look for signals from the first and last riders ☐

(c) Signal right, drive onto the other side of the road and overtake ☐
as fast as possible

(d) Flast your headlights and drive without crossing the white line ☐

What does this sign mean?

(a) Road closed ☐

(b) T junction ☐

(c) Turn right ☐

(d) Turn left ☐

You are driving your car and approaching this road sign.
You glance at your instrument panel (shown below).

Which one of the following should you do?

(a) Check your mirror and slow down ☐

(b) Look out for traffic joining from the left ☐

(c) Look out for traffic joining from the right ☐

(d) Check your mirror and signal right ☐

DRIVING STANDARDS AGENCY

Statement of Failure
Road Traffic Act 1988

An Executive Agency of the
Department of Transport

F 0000000

Name _____

has this day been examined and has failed to pass the test of competence to drive prescribed for the purposes of Section 89 of the Road Traffic Act 1988.

Test Centre _____

Date _____

Authorised by the Secretary of State to conduct tests.

Examiners have regard to the items listed below in deciding whether a candidate is competent to drive. The matters needing special attention are marked for your information and assistance and should be studied in detail. (see Note 1 overleaf)

1 ☐ Comply with the requirements of the eyesight test.

2 ☐ Know the Highway Code.

3 ☐ Take proper precautions before starting the engine.

4 ☐ Make proper use of accelerator, clutch, gears, footbrake, steering.

5 ☐ Move away safely, under control.

6 ☐ Stop the vehicle in an emergency promptly, under control, making proper use of front brake.

7 ☐ Reverse into a limited opening either to the right or left under control, with due regard for other road users.

8 ☐ Turn round by means of forward and reverse gears under control, with due regard for other road users.

9 ☐ Reverse park under control, with due regard for other road users.

10 ☐ Make effective use of mirror(s), take effective rear observations well before signalling, changing direction, slowing down or stopping.

11 ☐ Give signals where necessary, correctly, in good time.

12 ☐ Take prompt and appropriate action on all traffic signs, road markings, traffic lights, signals given by traffic controllers, other road users.

13 ☐ Exercise proper care in the use of speed.

14 ☐ Make progress by driving at a speed appropriate to the road and traffic conditions, avoiding undue hesitancy.

15 ☐ Act properly at road junctions:-
   -regulate speed correctly on approach;
   -take effective observation before emerging;
   -position the vehicle correctly before turning right, before turning left;
   -avoid cutting right hand corners.

16 ☐ Overtake, meet, cross path of, other vehicles safely.

17 ☐ Position the vehicle correctly during normal driving.

18 ☐ Allow adequate clearance to stationary vehicles.

19 ☐ Take appropriate action at pedestrian crossings.

20 ☐ Select a safe position for normal stops.

21 ☐ Show awareness and anticipation of the actions of pedestrians, cyclists, drivers.

Driving Examiners are NOT permitted to discuss details of the test   DL 24 (Rev 01/91)

Guidance Notes

1. More detailed advice about the test requirements and the items marked for your attention overleaf are given in the Driving Standards Agency's book "Your Driving Test". This book, published by HMSO (obtainable from bookshops) also contains the "Recommended Syllabus for Learning to Drive".

2. You should not only make a study of the book but ensure that you achieve basic competence in all aspects of the "Syllabus" before you take your next test.

3. The Department's manual "Driving" (obtainable from bookshops) and a study of the relevant rules in the Highway Code, will also be of help to you.

4. No further tests on a vehicle of the same Category can be undertaken until the expiry of one calendar month.

5. If you lose this form a duplicate may be obtained from the Driving Standards Agency, Regional Office.

Appeals

5. If you consider that your test was not properly conducted in accordance with the Regulations you may apply to a Court of Summary Jurisdiction acting for the Petty Sessional Division in which you reside (in Scotland to the Sheriff within whose jurisdiction you reside) which has the power to determine this point. If the Court find that the test was not properly conducted they may order a refund of the fee and authorise you to undergo a further test forthwith. (See Road Traffic Act 1988, Section 90).

6. You should note, however, that your right to apply to the Court under section 90 is strictly limited to the question of whether the test was properly conducted in accordance with the Regulations. **In particular the Court cannot question the merits of the examiner's decision.**

		Route Number

Weather Conditions

Brief description of candidate

Remarks

(2) x Six questions asked but candidate only answered one correctly. Five answered with 'I don't know'.

(20) x Forced pedestrian to step back on to the kerb when turning left into Park Road.

(20) x Followed closely behind two schoolboys on bikes who were fooling around and wobbling.

(20) x Candidate had to brake violently as a bus pulled away slowly from a bus stop.

Examiner's Signature

D 10 No

DL 24 No

Disability Tests - Including eyesight failures

Driver Number

Description of any adaption fitted

HC	X	(2)
PRE		3

^^ PED	X	(21)
CYC	X	(21)
DRI	X	(21)

Samples of Examiner's marking sheets with faults
recorded, and reminders of correct procedure
(see also pages 152/153)

Weather Conditions

Route Number

Brief description of candidate

Remarks

(5) x Candidate released handbrake before proper clutch control was achieved. Engine stalled several times and car rolled backwards out of control.

(5) D Candidate pulled out from behind a stationary van without checking for blind spot. This caused an oncoming Volvo to brake sharply and steer to avoid a collision.

(7) D Candidate looked all round but failed to see, and take appropriate action on, oncoming car. Caused it to brake and wait.

D 10 No

DL 24 No

Examiner's Signature

Driver Number

Disability Tests - Including eyesight failures

Description of any adaption fitted

MO	PRE	X	(5)
	CON	D	(5)

ES			6
	FR.BR		6

RV	CON		7
	OBS	D	(7)

	Route Number

Weather Conditions

Brief description of candidate

Remarks

⑧ x Candidate failed to select reverse gear correctly and rolled back into kerb.

⑧ D Candidate omitted observation to the left on final leg. Caused Ford Escort to brake sharply.

Examiner's Signature

D 10 No

DL 24 No

Disability Tests - Including eyesight failures

Driver Number

Description of any adaption fitted

TR	CON	X	
	OBS		D

Weather Conditions

Brief description of
candidate

Route
Number

Remarks

④ x Candidate never used the opportunity to get into top gear (4th)
throughout the drive, although we travelled at 30 mph on numerous
occasions.

④ x Candidate oversteered on a number of left turns. Twice allowed the
nearside rear wheel to drag over the kerb.

⑦ x Whilst reversing into side street, candidate slipped the clutch,
steered late and finished on the wrong side of the road.

D 10 No

DL 24 No

Examiner's Signature

Driver Number

Disability Tests - Including eyesight failures

Description of any adaption fitted

RV	CON	✗	⑦
	OBS		7

ACC		4
CL		4
G	X	④
F.BR		4
H.BR		4
ST	X	④

Part 6
After the test

Introduction to driving as a fine art

Once the driving test has been passed there still remain a set of
objectives to be achieved. In a way the responsibility for these is even
more demanding than for the previous ones. During your driver
training you will always have had your instructor or supervisor with
you to ensure that your training and practice were structured
towards achieving your final aim: that is, to pass the driving test, and
to become a driver for whom driving is a life skill. Your remaining
objectives will include the following, but they all point to the one
final objective: to drive safely with due regard to all other road users
every time you use the roads.

The objectives which you should strive to achieve in the next few
months and years are to be able to:

Drive safely and with confidence on urban and rural motorways,
entering, driving along and leaving them correctly and safely;

Overtake other traffic, making correct use of lanes, and making suitable
headway as necessary for road and traffic conditions at the time;

Drive to a standard acceptable to the Institute of Advanced Motorists, or
the RoSPA Advanced Drivers' Association;

Cope with a simulated or genuine advanced driving test, lasting at least
ninety minutes.

There are probably more myths about driving, driving tests and
driver training than any other sphere of activity – perhaps with the
exception of sex. One of the reasons for this is undoubtedly the veil
of secrecy which has always shrouded the subject of driving tests.
Hence the stories which abound concerning the unlucky people who

only failed their driving test because they wore an Arsenal tie with a Tottenham supporting examiner and the like. People only fail their driving test if they are not competent, or if, when asked to do so, they cannot demonstrate their competency. So what goes wrong after the test?

The theory is that if you drive during the test in such a way that you will not look as though you would cause an accident to happen, you will pass. This is true. Consequently, if you drive afterwards as though you were being tested you still shouldn't have any accidents. So why is it that people who have passed their driving tests fail to keep up this standard of accident-free driving afterwards?

There are two simple answers. First of all, people don't always keep up the same standard of safety as they did when they were taught to drive. The proof of this can be seen at every crossroad and junction. Watch the positioning that drivers take up when they turn right or left. And see how they fail to provide the safety gap between vehicles. Anyone who wants to study human behaviour at its most irrational could well write a thesis based on what drivers do to each other from the 'safety' and 'isolation' of their motor vehicles (hence the importance of eye-to-eye contact). The second factor is a more basic one. Virtually every driving lesson is based on an hourly module. Even when the driver training is conducted in an intensive fashion, it is rare for pupils to remain at the wheel for any length of time. Long-distance driving for learners is rare; rare and boring.

The reason it may seem boring is that the skills acquired on long-distance drives do not seem to fit into the pattern of driving tests. How often need one reverse, or do a turn in the road, on a long drive? Complaints are therefore often made that the driving test is artificial, and so are driving lessons. This is almost certainly true, and the reason is that few driving lessons have any specific purpose other than to prepare the pupil for the examiner and the test. Once the test has been passed, and the novelty has worn off, the purpose of each particular drive tends to override the theories of safe driving which had been instilled during the driving lessons. Sales reps are now concerned only with their next call. Football supporters are only interested in the match. Shoppers are thinking of their shopping lists and mums on the school run with losing their infants for a while! The skill of driving is relegated to the back of the mind.

It becomes, for the most part, a subconscious skill as simple as eating or walking.

The purpose of the post-test driving lessons is to make sure that this subconscious skill is a correct one. Driving on 'autopilot' is one way that it has been described. Once more we are back to the reason why most driving lessons rarely last longer than an hour or so: concentration. Until one can drive safely on 'autopilot', driving is an exhausting business.

So what is the ex-learner driver to do when, having passed the driving test, said thank you very much to your instructor, smiled at all the cards from friends which usually say such things as 'Congratulations, now learn to cope with traffic conditions like the rest of us', you find yourself with a licence, a motor car and somewhere to drive? The first thing is to realize that what the examiner said is true. You are now able and capable of driving on your own anywhere you wish. But logic also says that before you do anything fresh, or strange, you should think about it first so that you can tackle it in the safest possible manner. So what *are* the new things which are likely to get you worried?

Motorways Many new drivers worry about going on a motorway for the first time, especially when they have done so as a passenger on some very busy urban motorways.

Night driving Ideally your instructor would have tried to ensure that at least one of your lessons had taken place in the dark, but if not then the first night-time drive could make you feel apprehensive.

Adverse weather This covers everything from snow to fog and from ice to bright sunlight on wet roads. Previously during your driving lessons your instructor will have been able to select where you are going to drive, and in bad weather would have adjusted the lesson to suit your capabilities. Now for the first time you will have to make your own decision as to how, when and where to travel when the weather conditions are not so good.

Long-distance drives The prospect of driving a long way or staying at the wheel for a long time doesn't always seem so worrying until you are actually doing it. However, careful thought and a little bit of forward planning will make it more easy to cope with and safer for you, your passengers and all other road users.

Later on such things as caravan towing, continental touring, or even four-wheel driving and off-road driving may all be encountered. In every case there is no question that it is safer and cheaper to get some professional training beforehand rather than risk thinking it will be all right on the night.

Coupled with all of these new driving activities is the fact that ownership of a vehicle also carries with it some heavy responsibilities. Such things as excise duty licences, insurances, MoT car tests and vehicle servicing all need consideration, both from a financial point of view, and for legal reasons too. (See also 'legal requirements', pages 30–3 and 172.)

Advanced driving

Having passed their test, some drivers will now assume that they have all the necessary driving skills. In one sense they are right: they do *potentially* have all the requisite skills. But it is ridiculous to imagine, given the artificiality of the learner and test procedure, that this is all there is to the art of driving. There is no way that any responsible driver would accept this standard as good enough, and most new drivers do recognize that, having just passed their test, they now begin the process of really learning to drive. Many such drivers, having obtained a vehicle in which they can gain a little bit of experience, will think in terms of 'advanced driving', and apply to take a test with the Institute of Advanced Motorists or the Advanced Drivers Association of the RoSPA (formerly called the League of Safe Drivers).

The basic rules of driving are simple: keep a safe distance between you and everyone else. Never drive closer to the vehicle in front than a two-second gap. Never turn right (or left either) until you have confirmed that it is safe to do so. And never overtake until you are sure that you can regain your safety line without relying upon any other road user to slow down for you. Whether or not you decide to take an advanced driving test, if you apply these rules when you are driving and add the ingredient of experience you will become a better driver. Start thinking of what other people will do, and how it will involve you, and advanced driving is the result.

There are three stages to advanced driving which, if followed successfully to their logical conclusion, will make you one of the elite drivers who are in full control of their vehicle and themselves at all time. The stages are:

Smooth handling

Able to control the vehicle to such a degree that there is never any snatch or lack of smoothness when changing and selecting gear. The engine speed always matches the clutch speed which matches the gear box speed, which matches the driving wheel speed. This chain of smoothness can easily be felt when you are driving, or when you are a passenger in a vehicle. If the driver never causes any clutch drag during gear changes, a smoother ride is ensured.

Tyre grip

The second stage is to ensure that all four tyres hold the road with equal grip. Each tyre has a tread pattern about the size of your shoe. At rest, all four tyres grip the road by the same amount – that is one-quarter of the weight of the car lies evenly on each of the four tyres. If the car is moving the weight of the car is distributed according to the direction the car is travelling and whether there is any change of speed. A car which is accelerating will have more of its weight at the back of the car. If it is braking more weight will be placed on the front tyres. If the car moves sideways, such as when steering, then the weight is transferred to the left side when turning right, and the right side when turning to the left.

The overall effect of this on a moving car is that it is at its most stable when the car is travelling in a straight line and at a steady speed. If the speed increases the weight transfer begins. The greater the change, the more emphatic is the weight transfer. Now it can be seen why it is so important not to change speed and direction at the same time. If all the weight of the car is moved to the front through braking, and at the same time the weight is also transferred to one side through cornering, the overall effect is to finish up with all the weight of the car on one wheel: a definite recipe for disaster.

Therefore the skill of maintaining equal grip with all four tyres on the road surface becomes one which requires considerable attention.

What is needed is constant attention to the road ahead at all times so that the driver can plan the driving in order not to find a situation developing whereby the weight will be in the wrong place at the wrong time. Always brake early enough to avoid the danger of having to brake and steer at the same time.

Speed control on bends

The final skill of the three stages of advanced car control is the ability to look into and through bends in order to plan your driving in such a way that you always know when it is safe to accelerate and when you ought to be decelerating or braking instead. The way that this is done is to evaluate each bend as either an **opening** or a **closing bend** on approach. If it opens, then you can afford to maintain, or even increase, your speed. If it is a closing bend then it is essential to lose speed.

An example of a closing bend, when your *must* reduce speed

If the bend closes gradually then speed can be lost gradually; deceleration may be sufficient. If, however, it is closing sharply then it is essential that you lose speed now quickly enough to achieve the speed at which you know you can cope with the bend *irrespective of what may be happening around the corner.*

If you can gain these three skills then there is no doubt that you are an advanced driver. However, possession of any qualification – whether officially recognized or not – is no certainty that you will use it skilfully. The art of the advanced driver is skill with responsibility. And as you become more skilful, so you will realize even more the need to use that skill wisely, not only to keep your vehicle and its occupants safe, but also actively to help keep safe those who have placed their trust in others – including you.

Motorway driving

Although motorways have been with us for nearly forty years, the first time you drive on one can be quite nerve-racking, and you may wonder whether passing your test has really prepared you for this. But I assure you, when you received that pink pass certificate from your driving examiner this is precisely what it means: you now have a licence to drive anywhere you want, including on motorways, all on your own.

However, when you start anything new, whether it is tackling a motorway or towing a caravan, it may be advisable to get a professional to sit with you for that first time. Your own driving instructor will certainly be happy to help.

Motorways are in fact safer than other roads on a mile for mile basis. However, because speeds are generally much higher, a greater awareness of what is happening ahead (and behind) is called for. All traffic flows in the same direction, and there are no cross roads, right turns, roundabouts, pedestrians or similar hazards.

Perhaps the most important things to bear in mind when travelling on motorways are:

- lane discipline;
- separation distances;
- looking further ahead;
- constantly checking in your mirror.

The speed limits on motorways are simple and easy to remember:

- Cars, buses and motorcycles 70m.p.h.
- Car with caravans/trailers 60m.p.h.
- Lorries over 7.5 tonnes 60m.p.h.

There are no rules that say you have to drive at full speed all the way, and those who wish to may hold their speed down and remain in the left hand lane all the time, allowing faster traffic to get past.

That's the good bit. Unfortunately, many drivers treat motorways as their own private stretch of road. The problems start to arise when there is a speed differential between the various types of driver (in other words, it's not the vehicle that counts, it's the attitude of the driver). Where there are three lanes, the left-hand (inside) lane is intended for normal driving. It is also the lane by which you leave and join. The middle lane and the right-hand lane are intended only for those vehicles which are in the process of overtaking and are both called 'overtaking' lanes. The original aim of these lanes was to allow faster moving traffic to use them while they were in the process of overtaking. However, there is much more traffic using them than had ever been envisaged (by the 'Planners' that is; everyone else could see quite clearly how well used they would become). Consequently, there are so many lorries and cars using them that those who wish to travel faster than the average tend to stay in the middle and right-hand lanes all the time. Some drivers even stay in the right-hand lane the whole of their journey, apparently in the belief that because they are going as fast as they wish no one else shall pass.

It is here that problems begin. The national speed limit is 70m.p.h., but the 'average speed' of traffic using the right-hand lane of motorways can often be 80–85m.p.h. The result of this is that law-abiding motorists are forced to remain in the left-hand lane, where they are constantly harassed by lorries and traffic leaving and joining at various exits.

Motorways are divided into two basic types: rural and urban. The differences between them are enormous. A rural motorway joins two towns or parts of the country and allows traffic to travel in comfort at a leisurely speed and still make exceptionally good time. There are service areas every twenty or so miles which provide comfort stops and refuelling facilities. It is quite easy for drivers to pace themselves by allowing a reasonable gap between themselves and the traffic ahead with regular stops in which to relax.

Urban motorways (which are often extensions of the above) are shorter stretches of motorway which enable traffic to go across or round big cities and metropolitan conurbations by taking a longer route, but usually in much shorter times. However, the exits tend to be much closer together and many of the users of these motorways tend to join at one entrance and leave by the next exit. Consequently there is much more hustle and bustle on them and a lot of weaving in and out of lanes.

For the novice driver who wants to learn how to use a motorway for the first time, the best way is to join one from a major trunk road, where the motorway is merely an extension of the trunk road that they had previously been travelling on. This way there is no difference in the style of driving, except that now there are three lanes to use instead of two, and there are no traffic lights or roundabouts to contend with.

This is not always possible, however. If you live in any of the large metropolitan boroughs, quite often your first introduction to a motorway is to join something like the M6 at Birmingham, via Spaghetti Junction, or to join the M25 near London and find yourself travelling on what is alternately called the World's Largest Roundabout or London's Orbital Car Park.

The answer is to make the best of what you can. Take your time, plan your route and, above all, make sure you know where you are going and where you want to leave. Only overtake if you really need to, and be quite content to be the 'new boy', leaving it to the other drivers to race around making whatever speed they wish. Even if you take your time and don't drive any faster than you would on your favourite roads, you will still make better time and headway, simply because there are normally fewer holdups on motorways. If you drive like this you can almost enjoy yourself!

Once you have made a couple of shorter, local trips you then need to experiment a bit and see what it is like to change from one motorway to another. Once again, you will find this much easier than navigating in most cities and towns. The signs telling you which lane to take are much bigger, and much better planned, than any you will find in town. Keep your eyes well ahead, and know where you want to go. The rest is easy.

Motorway rules and regulations

Each motorway is built to a standard specification and comprises a pair of roads running parallel, separated by a central reservation. Each section of the motorway has either two, three or four lanes in each direction and a hard shoulder to the left. The hard shoulder is normally 10 feet wide, the three-lane carriageways 36 feet wide, and the central reservation is 13 feet wide with a physical barrier or safety fence between the two opposing overtaking lanes. Bridges have a normal minimum height clearance of 16ft 6in and there are telephones at one-mile intervals. The marker posts are at 110-feet intervals, which means there are 16 to the mile and 10 to the kilometre.

Motorway regulations commence at the very start of each motorway and continue through to each exit and the end of the motorway. The following are not allowed on motorways:

- Pedestrians;
- Animals;
- Learner drivers;
- Pedal cycles, and motor cycles less than 50cc;
- Invalid carriages;
- Agricultural vehicles;

and any slow moving vehicle except by permission.

Vehicles are not allowed to stop or park on any part of the motorway. If your vehicle has broken down, or is involved in an emergency, you should wait on the hard shoulder until your vehicle can be repaired or removed. If you do break down and are forced to use the hard shoulder, do not use your driver's door to leave the vehicle but use the passenger's door. Do not allow your passengers to remain in their seats or allow them to wait on the hard shoulder. Get them off the carriageway completely and the other side of the barrier if possible. Walk to the nearest telephone (it must be within half a mile) and then return to your vehicle to await assistance.

If your electrics are working, use your hazard flashing lights to warn other traffic and you can also place a warning triangle about 150 metres behind your vehicle (provided you have remembered to take it with you). All telephones are monitored by the police who control each stretch of motorway. If you are a member of the AA or RAC they will take your number and pass on your need for aid to the appropriate organization. If you are not a member, you will be put in touch with a garage and be required to pay whatever charges are made.

Over the past few years considerable publicity has been given to the problems associated with lone women drivers who have been attacked whilst sitting in their broken-down vehicles. To get this into perspective it is necessary to realize that the chance of being hit by another vehicle whilst you are broken down on the hard shoulder is about five thousand times greater than that of being attacked. Some authorities suggest that you should stay locked in your vehicle with your hazard lights on while waiting for a police patrol car. My advice is get out of the vehicle; if you are scared, keep a large spanner in your hand. You are much more at risk from dozy and dozing motorists, regrettably, than you ever are from opportunist attackers.

Driving on a motorway is something like flying. The easy bit is where you are just travelling along. The only concern is when you have to get on and get off. However, once you know the rules and abide by them then the whole thing becomes easier.

Joining the motorway

The first time you join an urban motorway the sheer speed and volume of the traffic can be quite overwhelming. But remember that the speed is only a problem if the traffic in your lane is not going at the same speed as you. Provided you are all moving at the same speed and keep reasonable distances from each other the basic rules of safety are being kept.

Use the slip road to build up your speed, and look over your right shoulder, and in your mirrors, for a suitable gap. When you see one, edge into it; but don't cross any hatch markings and try to keep that speed and distance, keeping in the left lane, until you have had time to acclimatize yourself to the speed and conditions on the motorway.

Driving along the motorway

When you are driving along the motorway only change lanes if it is necessary and suitable to overtake, and then pull back into the left lane again as soon as you can. Perhaps the only time to move out without actually overtaking is when you are approaching and passing a new entry to the motorway. By moving out to the centre lane you can make it easier for someone else to join the motorway from their slip road and acceleration lane, always provided that it is safe to do so.

Changing from one Motorway to another is simple; look for signs, both on overhead gantries and at the side of the road. And you will also notice that the white lane markings make it clear which lane you should be in for the route you intend to take.

Leaving the motorway

Leaving is also simple. You will know the numbered exit you are looking for; it will be signed by exit number and also road number one mile before you get there. Then half a mile later the names of the places where the roads go will also be given. Finally you will be given three hundred-metre countdown markers showing that you have 300, then 200 and then 100 metres before the start of the deceleration lane and you join the slip road. Once you are on the slip road make sure your speed is slow enough to cope with what is ahead, and look for your route at the roundabout or junction which will be coming up.

Remember the rules for motorway driving:

1 Keep your lane discipline;

2 Keep your speed consistent with the traffic flow;

3 Keep your distance, especially in bad weather;

4 Keep using your mirrors, especially before changing anything;

5 Keep looking for your route and exit signs;

6 Keep your eyes open for warning signs ahead;

7 Keep concentrating on the job in hand (driving);

8 Keep your cool; (and your temper too); and

9 Keep your eyes off things which don't concern you (especially traffic incidents which may be happening on the other side).

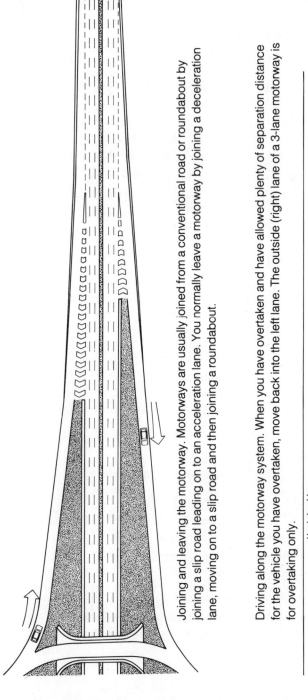

Joining and leaving the motorway. Motorways are usually joined from a conventional road or roundabout by joining a slip road leading on to an acceleration lane. You normally leave a motorway by joining a deceleration lane, moving on to a slip road and then joining a roundabout.

Driving along the motorway system. When you have overtaken and have allowed plenty of separation distance for the vehicle you have overtaken, move back into the left lane. The outside (right) lane of a 3-lane motorway is for overtaking only.

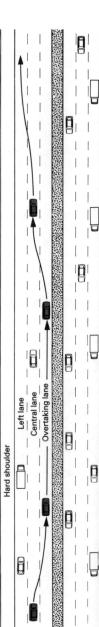

Hard shoulder

Left lane

Central lane

Overtaking lane

Skids, fog and flooding

Skids

Skidding is like marriage: most people get through their lives with it only ever happening to them once. Cynics amongst us sometimes add that like marriage – once is enough! Certainly the first time can be quite frightening, especially if it catches you by surprise.

A skid is caused by a vehicle travelling too fast and the driver suddenly wanting to change speed or direction quicker than the road surface and tyre grip will allow. The result is that the vehicle will continue to travel in the original direction, or somewhere between where the driver now wants it to go and that direction. The severity of the skid depends upon the forces acting on the vehicle at the time in relation to the surface and tyre conditions.

All of this says that if the driver is careful never to exceed the grip potential of his tyres on the road a skid can never occur. If a car is travelling at 30 m.p.h. on a good dry road surface a skid would be almost impossible. However, if the road surface is wet, the tyre tread inefficient, and the driver brakes excessively hard, it is quite possible for the car to lock all four wheels and skid. Similarly, if the driver were to turn the front wheels too quickly on very slippery conditions the car could continue in a straight line instead of turning.

In each case the driver is asking more from the tyres than they are capable of doing. So the answer to preventing a skid is always in the hands – and feet – of the driver.

There are three types of skid:

Those caused by **excessive braking** usually involve front-wheel skids, sometime four-wheel skids, because the tendency is to lock up the front wheels (where all the weight is transferred). Occasionally the rear wheels will lock as well. Depending upon the road surface at the time, the best advice is normally to release the brake, and then reapply it as soon as the wheels have started to revolve again. In practice the best advice for getting out of a skid is to stop doing what is causing the skid in the first place. However, many rally drivers – especially some of the Northern European ones – argue that on a dry road stopping distances are shorter when you keep the brakes pressed even in a skid. You must remember that if your front wheels are locked they cannot steer. It is also worth remembering from your emergency stop training that putting your clutch down increases the chances of skidding.

Skids caused by **excessive steering** mean that you must bring the wheels back to the 'straight ahead' position. If you lose your steering and continue to turn the wheel even more it will have little or no effect. However, if you do regain control your front wheels are facing a completely different and wrong direction. Skids caused by **excessive acceleration** usually result in wheel spin. This can often happen when a driver applies too much power to the driving wheels, quite often when starting off. Reducing the amount of acceleration will also remove the cause.

Usually, however, skids are the result of **a combination of excessive braking and steering**, either by braking and almost losing your grip followed by excessive steering; or steering at too tight an angle and then braking, causing the wheels to lock up and lose their grip. In either case the answer is always the same. **If you have room**, you can always survive. **Look** at where you want to go, release the footbrake, steer in the direction you want to go, and brake again gently if you can. However, if you haven't room, then survival is often dependent upon the other people and circumstances around you.

Skidding is of course an offence. You are out of control of your vehicle, and one of the requirements of the law is that you are in control of your vehicle at all times. Therefore you should always drive so as to avoid any danger of skidding.

There are two ways in which you can achieve this. Always drive well within the limits of your vehicle, its tyres, and the surface conditions which exist. And secondly take training in skid control, either on a skid pan or by using one of the numerous skid training vehicles which exist.

There are three basic kinds of skid training vehicle:

- Skid car consists of an H frame built around a car which lifts the wheels away from the ground creating a controlled loss of stability.

- Slide car is an adapted front-wheel-drive car which has rear wheels which also steer.

- Skid master is a specially adapted vehicle which allows the instructor to control the direction and braking on each of the four wheels.

Practice on a skid pan, or with a skid vehicle, is good fun even for those who are scared of skidding, and excellent training. What they do is to enable drivers to find out what sort of reactions they have, and how they would behave under skid conditions, but in a safely controlled environment.

The real benefit of such training however is that it allows you to learn *and practise* at slow speeds so that you never have the pressures and panic which occur when real skids happen.

Floods

Driving in floods and through fords can worry a driver, but if done carefully presents no problem. Drive slowly, but deliberately, through where you feel is the shallowest point in the water and you will find that your car will cope even if the water reaches half-way up the wheels. What you must avoid is rushing into any depth of water and creating a large bow wave which can easily swamp the electrics. It can help to stay in a low gear but slip the clutch as you keep the revs high in order to maintain a slow but steady speed through the water.

The final admonition after you have driven through successfully is to 'try and dry' your brakes. This is done by lightly braking as you drive on. This will dissipate any water that may have entered the drums or soaked the disc pads.

Fog code

When you are learning to drive, your instructor will try to take you out regardless of the weather conditions – heavy rain, snow, ice. After all, since you will have to cope with all weather conditions when you own your own car, isn't it best to do so for the first time with a qualified instructor alongside you?

However, there is one condition where driving instructors will not be keen to take you out, and this is during fog. So the very first time you drive in fog will probably be in the winter following your driving test.

This is when you will be feeling at your most vulnerable and you will probably worry more than you need.

Remember that in fog you must make every effort for other people to see *you* as well as for you to see *them*. Make sure you drive with dipped headlights on (main beams will reflect back onto you); use your rear fog lights too, if they are fitted. The benefit of these is that people a long way back will see them before they see your normal rear lights. Remember though to switch them off when the weather improves or if the traffic behind is travelling slowly with you.

Try to plan your route so that you avoid too many right turns. If you do have to turn right, make sure you don't cross the middle of the road too soon and that you have your indicators flashing. Look well into the road you are turning into before you commit yourself to start the turn.

If you have a passenger, get them to follow the kerb through their side window, using a hand lamp if necessary.

If you decide to abandon your car because the fog is too thick, don't just leave it at the side of a main road. Turn off into a side turning.

Remember that fog tends to blank out sound as well. If you are lost in the fog, keep your window open so that you can listen for other traffic.

Looking after your car

Vehicle characteristics

Although most car owners spend a great deal of time thinking about what kind of motor car they are going to buy next, the average learner driver has no say whatsoever about the choice of vehicle in which he learns to drive. Quite often learners have ideas of what sort of car they would like to drive or own, and some insist on paying more attention to the type of vehicle on which they will learn, than to the qualifications of their instructor. Nevertheless, most pupils regard the car they are learning to drive in as something which is determined for them.

They do not therefore take into account such things as the differences between front-wheel-driven cars, rear-wheel-driven cars or even four-wheel drive. At this early stage of learning, this is probably just as well, as any choice they made would probably be based on a wrong premise anyway. These days most small cars tend to be front-wheel drive. When the Mini was first launched at the end of the 1950s many experienced motorists found themselves in a dilemma, not being too sure how to handle a new style of vehicle with completely different handling characteristics from any they had encountered before. Since then, whole generations of new drivers have learned to drive, passed their tests and then bought front-wheel-drive vehicles without being aware that there is any other form of transmission chain.

It is almost as if front-wheel drive is natural and all other forms of transmission can be discounted, from a learning to drive point of view. Yet all learner drivers should ask their instructor what kind of transmission the car on which they are learning has, and how the handling characteristics vary from others. At the moment, four-wheel drive is rare amongst smaller cars, yet this is becoming an option which most manufacturers are giving to the customers. Although the differences in handling on good dry roads are minimal, there is sufficient variation when the road surfaces offer less grip for every learner to be taught what they are.

In general, a front-wheel-driven car will grip the road better on bends than one which is driven by the rear wheels. And a four-wheel-drive car will grip the road even better than either of the above. However, it is only at speed that such differences become a matter of urgent concern. Sufficient for this book to say that every driver is to be encouraged to find a suitable instructor who can demonstrate and explain the benefits of each system, after the driving test has been passed and the client is considering a change of vehicle.

One other potential difficulty for the learner driver is to understand and recognize the different ways in which a vehicle can handle according to the load being carried. A small car with only an instructor and pupil on board will brake and stop much more quickly than one which has a whole family and their luggage on board. A factor which has spelt disaster for many holidays. No one had explained that the weight carried on any vehicle has a marked effect on the way it handles at speed and especially whilst braking.

Tyre tread depth, tyre pressures, the tracking of the front wheels, the effect of weight transfer when braking, accelerating or cornering all need to be understood in practice as well as in theory.

Imagine that your vehicle is a tin box with a wheel at each corner. If you put your load in the middle the tyres will grip the road surface equally. If you increase your speed by acceleration the weight is transferred temporarily to the rear wheels, and less grip is exerted by the front wheels. If the brakes are applied, the weight is transferred to the front wheels and the grip on the rear tyres is much less.

When cornering the weight, and therefore the grip on the tyres, is transferred towards the wheels on the outside of the arc being steered. It can therefore be seen that when a vehicle is both braking and cornering all the weight is transferred to one wheel only – the front tyre on the outside of the curve. If for any reason that tyre loses any of its grip the vehicle would become totally unstable. It is for this reason that you should never brake and turn the wheel at the same time. For similar reasons it is also inadvisable to accelerate and steer at the same time.

Certain types of vehicle handle with more difficulty under such circumstances than others. For instance, trucks or vans with high sides are much less stable than smaller cars with a lower wind resistance, especially in strong winds. This wind can also affect even small cars at times. Perhaps the most notable occasion is when you are passing a high-sided vehicle at speed and in high winds. Your own vehicle can be badly affected after you have passed the truck. It is important to ensure that you keep a strong grip on the steering wheel at such times. And of course bear in mind the fact that cyclists and motor cyclists can easily be blown about by strong winds.

Legal requirements for your vehicle and maintenance

Every driver must be aware of his responsibilities in relation to the vehicle he is driving. Your instructor should know these requirements and you would be well advised to check these matters with him while you are learning to drive. During your lessons, get into the habit of familiarizing yourself with your instructor's car and its upkeep. Before you start each lesson, check the tyres, the lights, the general condition of the car, and satisfy yourself that you know what sort of things to look out for.

The law requires that all vehicles over three years old comply with what is known as the MoT car test. This ensures that, at the time the test is conducted, the vehicle is safe with regard to lighting, brakes, steering, exhaust and similar items. You, as the car owner or the driver, will be held legally responsible if your vehicle has any potential weakness that may spell danger. The financial benefits to yourself of looking after your vehicle are also important.

If you are unable to service the car yourself, then you will need to arrange for a local garage, or mechanic, who will do it for you. On the other hand, even the least mechanical of drivers would benefit from understanding and appreciating what makes a car move (and stop). There are two essential items needed: fuel, usually petrol or diesel; and electrical energy usually supplied through a battery and an alternator.

If you prefer to read and learn from books, ask at your local library for a good helpful book at your level of interest and knowledge. Or take your time browsing in your local bookshop. Better still is the opportunity to explore the courses at your local evening classes. You don't need to get your hands dirty to learn how a car works. But the knowledge you do learn might save you some embarrassment on a wet and windy night on some deserted country lane.

Let us look at these facets of a car which require regular attention.

Steering As you can see, there are a number of links in the chain between the steering wheel and the tyres. Each of these links is a place where wear can take place. Ask your instructor to let you feel the steering wheel while standing outside the car. As you steer the wheel gently you can see that the tyres will try to move slightly; you can feel that a certain amount of play is possible. Let your instructor demonstrate exactly how much flexibility should be allowed to exist before you need to consult a garage.

The brakes, footbrake and parking brake When you press the footbrake pedal you should be aware of quite strong resistance to your foot. It is this resistance which tells you that the fluid in the brake pipes is doing its job. When you apply the handbrake you should count the number of clicks it takes to lock it into place. This is the only time that you should ever apply the handbrake without pressing in the ratchet. Six clicks tell you that the brake is correctly applied and adjusted; more than seven or eight clicks is too much and an indication that brake adjustment may be necessary. Not all cars use hand-operated systems for parking brakes however. In some vehicles the parking brake is applied by your left foot – it can't be the right one, which is needed for the footbrake. The parking brake in this instance is released by a lever operated by the right hand (to enable you to manage good clutch control).

The tyres should be inspected visually, and occasionally by running your hand all round the tread and side walls to ensure there are no bumps, nails or other uninvited guests in them. Again, ask your instructor to allow you to learn how to check the tyre pressures – and what are the correct times to do them. A forecourt facilities lesson by a professional instructor may be one of the most important ones you'll ever have. Find out where the tyre pressures for your vehicle can be found, and how to ensure they are correct.

When driving, your life depends upon the grip of four pieces of rubber the size of your shoe (the wheel, of course, is considerably larger than a shoe; but only a fraction of the wheel is in contact with the road at one time). It is therefore important that you realize the need for your own involvement in their condition.

A number of drivers, especially those who have recently passed their driving tests, take only a limited interest in the mechanics of their vehicle. They try to find a convenient friend or relative who is interested in engines and allow them to keep an eye on what goes on under the bonnet. That may be sufficient. Engines on the whole are fairly reliable, if dirty, things. It is possible to clock up more than 100,000 miles without an engine ever needing a major overhaul. Mind you, a thousand miles in an engine's life is usually the equivalent of a year in the life of a human. So a car with more than 50,000 miles on the clock is likely to take its time getting up hills, and be quite grateful for any breathing spells offered to it after any exertion. Once a car has reached 70,000 miles it may still keep going, but it does have a pension book in its glove pocket.

Tyres, however, are a different matter entirely. Each tyre is composed of a chunk of rubber which is gradually wearing away. The dust you see on the roads is in part composed of bits of rubber which have been left behind by every tyre which has driven over that road. The life of a tyre is unpredictable. It depends a lot on the way the car is driven (see pages 179–81 for further information, and especially the section on skids, page 167) and the condition of the roads used. But they all wear away. Sometimes a tyre will last 25–40,000 miles. Occasionally it will wear out after only 5,000 miles or so. You as a driver are not only relying totally upon your tyres, you are completely responsible for them too.

Look after them and they may look after you. Neglect them and they will certainly let you down. And when they do, it will never be in a garage on a nice warm day. One of the most frightening things you may ever have to do is to change an offside wheel on the hard shoulder of an unlit motorway in the teeth of a gale late one winter night.

Wheel changing

Although the whole procedure only takes about ten minutes, changing a wheel is one of the most frustrating and infuriating of all

When changing a wheel, make sure that the jack is secure and has a firm base – even a book will help.

driving tasks. The procedure is in fact simple and straightforward, but only if you follow the correct sequence. Try to change a wheel whilst the car is in a safe position. If it is on the driver's side and you are on a busy motorway, don't change a wheel by yourself. This is the best time to get the AA or RAC to help you.

If the traffic conditions are safe, follow this sequence:

1 Remove the wheel trims or hub caps.
2 Make sure that the brakes are firmly applied (if it is a rear wheel, also use chocks on the other wheels to prevent them rolling).
3 Loosen the wheel nuts before you raise the jack (only ¼ turn).
4 Jack up the car using the jack provided and making sure it is correctly positioned according to the handbook.
5 Raise the wheel until it is free from the ground (about an inch is enough).
6 Undo all the wheel nuts and put them in a safe place (inside the hub cap or wheel trim is best).
7 Have the spare wheel ready, beside you, remove the old wheel and replace immediately with the spare. Put the wheel nuts on finger tight straight away. Make sure the curved ends of the nuts are facing the wheel.
8 Lower the jack sufficiently to hold the tyre to the ground. Tighten up the wheel nuts in a diagonal sequence.
9 Lower and remove the jack and pack away the old wheel. Remember to have it repaired or replaced as soon as possible. You can only change a wheel if the spare is correctly inflated and in good condition.

A few extra pointers are worth bearing in mind. Always make sure that the ground underneath the jack is solid. If necessary you may have to use a large piece of wood underneath to prevent the jack sinking into the road. Don't ever crawl underneath any car which is supported by a jack. And if changing a wheel involves you putting your body in the line of any other traffic don't do it. Either get someone to help you and control the traffic, or call out the AA, RAC or National Breakdown services.

Check your tyre pressures at least once a week. Try to use the same service station (and the same gauge, tested on cold tyres) each time.

That way you are not likely to be misled by a false or strange reading. Read the manufacturer's handbook for your car to see what the pressures should be for your particular vehicle and the tyres used. Make sure that you have equal pressure in each pair, and never allow any one tyre to develop a slow leak that you feel you can 'cope with by topping up a few pounds every Sunday'. If you have a leak you have a potential danger. If it is a slow leak it may be a slow danger. If you lose pressure consistently, have the tyre checked or changed. Don't put it in the boot and keep it as your spare. When you do need it it is sure to be flat.

Finally, when you do visit a tyre-fitting establishment, have a good look at the tyres which have just been discarded by other motorists who have now fitted new ones. Look at the state of some of them and be grateful that they have now been changed; but remember that many of the vehicles on the road coming towards you at speed on wet and greasy roads have tyres just like these.

Lights and reflectors Having checked to see that all the glass/plastic lenses are not cracked, broken or discoloured, ask the instructor to sit in the car and switch on each of the lights in turn so that you can confirm they are lit, and in the case of indicators flash correctly. Once more your instructor can help you learn how to check that your lights, especially brake lights, are working when you are in your own car. Remember that apart from being illegal, faulty brake lights make it much more likely that following traffic may run into the back of you. Many cars these days, perhaps including your instructor's car, have warnings on the fascia panel to tell you when any of the car's lights are not working, or even when any of the doors or boot lid are not correctly fastened. However, when you buy your own first car you may still have to learn how to check them manually.

Exhaust

While you are still outside the car, ask your instructor to switch on the engine so that you can see the exhaust. Ideally, although you should be able to feel, or even smell, it you ought not to be able to see it. If you can, and especially if the smoke is black or oily, it may be that there is something wrong with the engine. If you or your instructor are 'green minded' you can also use a piece of cotton waste to test that the smoke is not polluting the atmosphere.

Wipers, washers, seatbelts, horn

Also ask your instructor to use the wipers and windscreen washers to show that they are working correctly and efficiently. This especially applies to windscreen wiper blades which when worn are worse than useless. They can smear the screen and make it difficult to see properly ahead. Don't forget the rear window – if it is fitted with a wiper – and check that the rear washers work and the container has enough water. Once you are sitting in the car, but before you move off, you should also check that your seat belt, and all the other belts in the car, are effective and clean, and that all of the mirrors are working, correctly positioned, and clean. The only other items which require specific attention each time before you drive off are the speedometer, other gauges, warning lights and finally the horn. This latter item presents a slight problem. The law does not allow you to sound the horn whilst stationary; and you ought not to sound it unnecessarily whilst on the move. However, it will be too late to find out it doesn't work if it is needed urgently, so get your instructor to show you exactly how it works and what sort of noise it makes.

All of the above items are automatically tested as part of the MoT car inspection system. Once you own your own car and it becomes three years old it will require an MoT certificate before it is allowed on the road. The best way to learn how to avoid unnecessary bills, or having to keep your car off the road until it can pass the MoT test, is to know beforehand what items are to be tested and what these requirements are. Early identification of faults not only saves money on expensive repair bills, it can also save lives.

As the driver of any car, you are responsible for all aspects of it; even the way it is loaded. Never be tempted to carry more passengers, or a greater load, than that recommended by the vehicle's manufacturers. And if you have any children in the car at any time, remember that you are also completely and legally responsible to see that they comply with the laws regarding seat belts and safety seats. Adults, and children over 14 in passenger seats, are legally responsible for their own seat belts; but you as a responsible driver can always refuse to carry them if they do not comply with the law with regard to seat belt wearing.

Green for go carefully, or how to save wear and tear

The benefit of driving in such a way as to save fuel is not just a matter of jumping upon the 'green bandwagon'. Good drivers have always driven cautiously and tried to save fuel because it makes good economic sense. The added benefits are that when you make a conscious effort to conserve energy you will also find that you are naturally driving more safely too.

Let me explain. The pedal on the right is called an accelerator. It ought to be called a 'tap' because that is exactly what it is – a petrol tap. Except that unlike your kitchen or bathroom tap it doesn't really switch off, except when the engine is switched off too. When the engine is running the tap is open all the time. When you put your foot down all you are doing is opening the tap even more. So fuel economy starts from the moment you sit behind the wheel. Don't switch on the engine and leave it running for some time just to get the engine warmed up. You are wasting an enormous amount of fuel whilst you are doing this, especially if the choke is being used as well. Ideally you should start the engine and get the car moving as soon as you can. The second thing you should do to conserve fuel is to realize that every piece of acceleration should be used. Therefore each time you brake you are effectively wasting the fuel you have just used. Always press the accelerator pedal gently, increase speed by all means, but don't do it by shoving your foot hard down on the pedal. Try to think of it as if you were allowing money to flow through the floorboards.

To avoid braking is not as silly as it sounds. This simply means not rushing into things at high speed, but reading the road ahead of you so that you are already decelerating early enough to avoid having to brake harshly. Hard braking has two penalties attached to it. One is the waste of the fuel you used getting to that speed only to lose it through braking; the second is the extra fuel you will now use to get back up to that speed again.

The most expensive journeys are those which consist of short sharp shopping trips. The most economic are those where you are driving a reasonable distance at a steady speed without any excessive acceleration or harsh braking. When you buy a new car you will often see that various fuel consumption figures are quoted for it, including one at a simulated 56 m.p.h. Quite often this is the one that shows the greatest fuel economy for that particular model.

However, there is one bonus for you – apart from saving costs – and that is through reading the road ahead you will become a much safer and more considerate driver, not only to your own vehicle but also towards other road users. By becoming more aware of other road users and what they are doing, you will always find yourself in a hold-back position, decelerating instead of leaving it later so that braking becomes necessary and occasionally urgent.

The same system of car control which is used in Defensive Driving Training systems and in High Performance Training is also used in 'Economy Driving Training'. To avoid using the brakes you will need to work very hard at your forward planning, to make sure that whenever a situation ahead is building up into one that is likely to require you to slow down you will already have started to do so by not accelerating, then by decelerating, and only using the brakes as a last resort.

You will not only be looking to see what other road users are doing, you will also be concerned with the general geography and topography of the journey ahead of you. By continuously scanning the road ahead you can plan for changes in gradient to avoid late gear changing and sudden rushes which result in flooring the loud pedal as hard as you can. Overtaking is a supreme example of how good, safe driving techniques can be used intelligently and without excessive use of the accelerator. A simple rule when overtaking is never to try to start to overtake from immediately behind the vehicle you wish to pass. Stay well back, then pick up speed from well behind it and gradually increase your speed until you are able to get past and back into your side of the road without needing to brake afterwards. Needless to say you should always take the greatest of care when overtaking. No point in saving ten per cent of petrol money if your insurance claims rocket skywards.

It is much easier to save fuel if your car is in good condition. Ensure that it is regularly serviced, that all the air filters, plugs or other bits and pieces have been checked and replaced where necessary. Similarly, ensure that tyres are at the correct pressures and are properly tracked. Unnecessary tyre wear is just as expensive as fuel.

You might be tempted to try 'racing lines' on bends in order to avoid slowing down for them, but this is a false economy and has no place in safe driving. Apart from anything else, if you are in the wrong position on a bend, suddenly needing to get back into your own side can be more expensive than just wear and tear on brakes and tyres. Instead of this try to plan your driving in ten-minute stages. Every now and again while on a long run see if you can manage to hold your speed steady for as long as possible, keeping your right foot as lightly as you can on the accelerator pedal and not using the brake at all. After a while you'll find that you can do this for quite long stretches, and although the 'ten-minute' test isn't always possible, as soon as you have had to use the accelerator or brake pedal or carried out a gear change try again.

Most cars these days have five gears as standard. Try to get into top gear, whether it is fourth or fifth, as soon as you can with regard to the comfort of the engine. The object of giving more gears to modern cars – some are now using seven – is to enable the engine to be run at maximum torque within as small a range of engine speed as possible, usually between 2,500 and 3,500 revolutions.

Finally, on the all important subject of fuel economy, there are two ways in which you can help yourself. The first is to buy – and use – a vehicle which uses unleaded petrol where you can. The second is to think very seriously about the benefits of using diesel, especially a 'turbo diesel' engined car. The economical benefits of these can be remarkable.

Automatics

Throughout the whole of this book the assumption has been that you are learning to drive in a motor car fitted with conventional clutch and gears. However, many learners – for a whole host of reasons – decide to learn to drive using a motor car fitted with automatic gears, and no clutch. 95 per cent of the material in this book still applies exactly the same to these learners as to all others. If you think of it, when you see a car approaching, you don't wonder which gear the driver is using, let alone if it is an automatic.

The Driving Examiner who will sit alongside you to give you your certificate of competence isn't bothered either. His only concern is whether you are competent or not. And therefore this is what you and your instructor will also be most concerned about. If you buy a camera these days you can opt for a purely manual one, one that is completely automatic, or one which you can choose to operate manually or automatically. In a way an automatic car is similar to the third of these. Driving an automatic car does not relieve you of any responsibility to other road users, it merely saves you the problem of wondering about which gear you should be in and using the clutch.

So what are the differences? First of all, you will notice there are only two pedals instead of three. But the two pedals are still both operated by the right foot. If you are driving an automatic you can tuck your left foot under the seat out of the way, unless the foot brake pedal is so large that it can be used by both feet, but only do this in a way your instructor will demonstrate. The second immediate difference is that the gear lever will be a different shape and have strange letters on it:

D for drive, this is your normal driving position;

N for neutral, for when you are parked or stationary;

R for reverse (you still have to go backwards occasionally).

Other letters will tell you that you can P for park; or L for lock or hold the gear you are in to prevent it changing; and variations on the theme of D1, D2 and so on. Again this is so that you can actually use the gears on the automatic just as you can on a normal geared car, but without the bother of a clutch.

This leads to the only drawback (for a learner driver) on the use of an automatic car for driving lessons. Throughout this book attention has been drawn to the benefits which accrue to the driver when the skill of clutch control has been totally mastered. You can't have clutch control in an automatic, so you have to be just as skilful at the art of slow moving by using the accelerator and brake pedals, **very gently**, instead. This is one of the few occasions when your instructor will allow you to use both your feet on the pedals instead of just the right foot.

There are two other important items which you need to take into account when driving, or learning to drive, in an automatic car. The first and most important is that when you leave the accelerator pedal alone you will still have a certain amount of 'creep' built in. That is, the engine speed is so adjusted that without putting your foot on the pedal the engine would still move the car very, very slowly. When you are using choke to help your start first thing (and it may be an automatic choke as well), this engine speed can be faster than you want or expect. Therefore you must always apply the parking brake whenever you stop, even if only for a brief moment or two. Your handbrake is your lifeline in these circumstances.

The second 'extra' is that device known as 'kickdown'. When you need extra speed in your manually geared car you can always drop down a gear or two in order to give you the extra acceleration that you need. In an automatic, if you kick the accelerator pedal down to the floor you will get the same effect even if you didn't want it at that time. So you need to treat your pedals with just as much care as the driver with the clutch control.

In terms of driving skill, there is nothing to choose between driving a manual car or one fitted with automatic transmission. The choice of which to learn on is up to you. However, you must bear in mind that if you do pass a driving test on an automatic car, you will be given a Category By licence and you must take another test on a manual before being allowed to drive a manual vehicle unsupervised. If you pass on a manual car then you are automatically covered for automatics as well.

If you are thinking of learning and are not sure whether to learn on an automatic or manual drive car, talk it over with your instructor and be guided by what he tells you. Remember that you will be tested much more on your ability to control situations than on the gears.

Driving through the year: hazards and warnings

Over the past twenty years or more the press and others have highlighted the annual death rate over the Christmas holiday period, yet this is the one season of the year when deaths caused by driving and riding are at a minimum, despite the excessive amount of alcohol which is undoubtedly imbibed at the festive season. Look out of your window any Christmas afternoon and the thing which will surprise you is the obvious lack of any traffic on the roads. Drive down a motorway on Christmas night and you will find you can actually circumnavigate the M25 without a single hold-up!

If you really want to find a holiday period when drivers apparently go completely mad, try Easter for size. Starting on late Maundy Thursday afternoon, the early holidaymakers join the caravanserai to the coast. The roads are littered with shattered bits of caravans; and luggage mingles with broken crockery, bedding and children's toys. In fact the whole of the summer period sees an increase in all vehicle-related accidents, whether connected with drink or not.

The reasons are obvious and predictable. Over 40 per cent of the annual death rate of children on the roads occurs between May and August consistently each year. There are more opportunities for youngsters to be on the road, and many of the accidents happen in their own streets. Eighty per cent of all child accidents occur within half a mile of their homes, with children in the 10 to 14 age group being the most vulnerable.

Another feature which affects the summer statistics is the fact that many drivers put their cars under much greater stress during their annual holiday trip. A car which is normally used for Sunday afternoon trips and commuting to the station is suddenly loaded to the brim with all the family, and then their luggage is added, almost as an afterthought, quite often secured with rope and string to a temporary roof-rack. Then, instead of a leisurely drive to the station or school, a five-hundred-mile motorway jaunt is undertaken. No wonder something gives at the first opportunity. It could be the roof-track, or the pistons; it might even be the brakes or steering, which will only give way at a crucial moment when overstressed.

A final check of your vehicle before a long journey is always a sensible thing to do. Better still, arrange a service and tune-up with a professional garage so that someone else has given everything a thoroughly keen scrutiny.

Driving on unfamiliar roads is always stressful, especially to the new driver; but driving on roads which you used to know can be worse. If you have never been to a town before you and your family can all sign-spot for you. But if you have vague memories of going 'this way', you can easily find that a new one-way system will thwart you.

Reading the map, and planning the route, are essential items for every long-distance journey, but even more so for the new driver, or the new car. Above all, when planning a long journey, whether for the first time or not, allow yourself plenty of time for comfort stops. If you are taking children with you, allow even more time, and make sure that they have plenty to occupy themselves on the way.

In winter the problems faced by drivers on long journeys are somewhat different. Ice, rain, snow, hail, fog and even winter sun all create their own difficulties (see pages 167–70 for more on bad weather conditions). Perhaps the most obvious one is falling snow: the main problem is usually one of visibility, so switch on your lights, not so much to see as to be seen. Drive slowly in the highest gear that the engine will accept. And do every manoeuvre gently and slowly, avoiding harsh movements on the steering wheel or brakes.

Ice can be worse because quite often the first time you know about it is when you feel it. 'Black ice' is the term used by motorists when they refer to ice that cannot be seen. If the weather is cold you ought to be looking for symptoms of icy roads all the time. Do everything deliberately, and more slowly. If you have to brake, avoid long sustained applications of the brake; instead use short, gentle and smooth braking whenever you need to slow down.

Rain is not restricted to the winter of course, but remember that quite often you are legally required to put on your headlights when it is raining and visibility is limited. Avoid hitting large pools of water at speed, not only because it can splash other road users but because you never know how deep a pothole might be.

Hail can also play havoc with your road grip and with your visibility. Remember too that pedestrians and cyclists often duck their heads and hope for the best when it is raining or hailing.

Finally, the sun in winter has its own particular hazard, especially early in the morning or late afternoon when it is low in the sky and can blind you to other people, or make other road users unaware of you.

Drink and driving

Although everyone knows that driving and drinking do not mix, and although people now regard those who do it and get caught as dangerous criminals who deserve to lose their licence, nevertheless too many drivers consider that they themselves are quite capable of drinking just a little bit and still driving safely. Their argument is based on the principle of 'I know what *I* am doing'.

Unfortunately, they don't. They know they can manipulate their own car, provided nothing happens to break into the autopilot system. When this happens, their reflexes are just too slow, and death, injury and destruction are the result. For the benefit of those who think they know all the facts, here they are to refresh their memories. Those who may need ammunition to counter the argument of 'I know what I'm doing' can use them too.

The legal limit for anyone is 35 microgrammes of alcohol for 100 millilitres of breath. But this is not the maximum that anyone can absorb without being affected. Those who are not used to drinking will be affected long before that limit is reached. Just one small sherry can be too much for some drivers, therefore the only safe limit is 'nil'.

In a third of all road accidents alcohol plays a significant part. No one can drive better 'after a drink'; you may be more relaxed, but you will also be less observant, and your reactions are much slower. Your abilities to analyse situations and make correct judgements are considerably reduced. Your eyes take longer to focus, especially in changing patterns of light at night. And your attention span is shortened and more likely to wander.

No wonder that driving and drinking do not mix; and no wonder that the penalties – whether penal, financial or moral – are so great. If you drink and drive, it isn't a question of if you win or lose; it is a question of how much you lose. Do you just lose your licence? Or your job as well? If your job, will you lose your income and your house? Or do you simply lose all your self-respect knowing that the loss of another person's life is entirely your responsibility?

Protect your children in the car

For many years parents have assumed that their children are safe while travelling in the rear seats of their vehicles. Even the least safety conscious of people are aware of the sheer stupidity (and illegality) of sitting a baby on mother's lap and assuming that in the event of a bump the child will come to no harm. Nowadays, rear seat belts are a legal requirement on all cars first used from April 1987. And all drivers and passengers are legally required to be secured in belts or restraints according to age and fitment.

From birth to about twelve months the best method of securing your infant is by use of a rear-facing infant carrier. These special fittings are held in place by an adult seat belt and face backwards, providing extra support. This type of carrier is easily removable from the car and can be used as a conventional baby seat.

From eight to twelve months to about four years children should be restrained in special child safety seats. In these seats the child is kept restrained with added side support for sleeping. Many seats are fixed to the car structure by straps top and bottom. Others are held in place by the adult seat belt.

From the age of four years upwards, a child harness and booster cushion will enable most ages and sizes to be catered for. A child harness normally consists of two shoulder straps and lap belt. They can either be attached directly to the existing anchorages or to a bar running across the width of the vehicle. Booster cushions should be used for older children, which will lift them from the normal seat to fit them better into the seat belt, and at the same time allow them a better view. It must be noted that these booster cushions are specially made with hooks and straps to hold them into place. Ordinary cushions should not be used in their place. Apart from any other reason, they could allow a child to slide under the belt and thereby be strangled.

Fitting of all the above extra belts and seats is fairly easy. Provided the seat belt fixings are already there it is easy for the average car-owning parents to fit the child restraints themselves. However, if there is any doubt, get the advice of your local garage. Cost is minimal in every case. If the security of your child can be measured by a sum of less than fifty pounds, then the alternative doesn't bear thinking about.

Accidents and basic first aid

Every car should carry a first aid kit (though not on the back parcel shelf where it becomes a potential lethal weapon in the event of any dramatic stop). This is not enough in itself, however. A first aid kit needs someone to use it when needed, and even the briefest of training will stand you in good stead should an emergency situation arise.

The St John Ambulance Brigade and the Red Cross both give occasional short courses in the basic elements of first aid which probably last for less than three hours and yet, if used intelligently and quickly enough, can save a life. It is sometimes possible to arrange training through your business or works facilities. First aid is precisely what it says; an initial aid to the injured whilst they are awaiting the arrival of medical or ambulance facilities.

If you are first at the scene of a road traffic accident, there are certain basic precautions you can take which are essential, even if you are not skilled in first aid and cannot help the injured in any specific way.

First of all, warn any approaching traffic. It is important to prevent any other road user becoming involved, thereby exacerbating the situation. Try to organize any others who are uninjured to assist with this warning, but do make sure that these warnings (whether car warning lights, hand signals or temporary signs) are deployed sufficiently far from the scene of the accident. Too close, and they could create problems in themselves.

Next try to assess the extent of the damage. How many vehicles? Are there any people still inside any of them? Are they trapped, or can they be helped out? Is there any danger from fire? Even if not, still impose a strict 'No smoking' rule within the vicinity. If you have time and can do so, isolate any vehicles' batteries that may be accessible. Certainly you should switch off any ignition switches which have been left on.

The next step is to find out if anyone present has had any first aid or medical training. If not, it may well be that you will have to make the decision to help as best you can, or to leave the injured until professional help arrives. Finally, of course, you need to summon help. If you are the only person at the scene, the sooner you do this the sooner you can get on with the other things that have to be done. However, if there are a number of people at the scene you may well be the one who has to galvanize the others into action. When you call the emergency services on the telephone use the standard 999 service. Ask for Police first, and allow them to alert the Ambulance and Fire Brigade services if necessary.

Use a public call box if you can see one, otherwise try the nearest house or shop. The cost of the call is free, but if you ask the householder to make the call for you ensure that they know exactly what has happened. If you make the call yourself make sure you know the exact location and the scale of the accident. If necessary send another motorist in search of the nearest call box or house. Be prepared to duplicate this action if necessary, rather than risk not getting through.

If you are the one who is to administer first aid, here are some basic rules.

Shock Many people involved in accidents suffer from shock. If this is apparent, try to get the victims to sit or lie down and get them to loosen their clothing. Do not offer them drinks or cigarettes.

The first thing to check is **breath.** Are they breathing? If not you need to do something drastic to get them to do so. Clear any air passages if they are obstructed. Ensure their tongues are not blocking an airway, and lay them down in the recovery position. If you don't know or understand the term it might be better if you are the one waving down oncoming traffic.

Next look for **blood.** Any minor wounds should be exposed and the bleeding controlled with an adhesive dressing or similar. If you have anything to clean the wounds with, such as antiseptic lotion or cream, use this. If the bleeding is considerable try to raise that part of the body to a higher position and use a pad to restrict the bleeding (see also broken bones). If the bleeding is very severe, try to staunch the flow with a pad of any firm material; even an oily rag is better than nothing at all. If the blood is flowing that fast germs aren't going to get in, and you need to prevent too much loss of blood.

Finally look for **broken bones.** If there is any danger of fracture or dislocation do not move the person at all (with the possible exception of danger from fire). Do not try any first aid at all, unless you are a trained nurse or have had considerable medical training – in which case the aid you would be offering would not come under the heading of first aid.

If you have any injured, concussed or dazed occupants who are capable of talking, ask them if they have any pain, or if they suffer from any illness or allergy. If they do, write it down (in case they lose consciousness before help arrives), and pin it to their clothing. Look also for any 'Medical Information Tags' which are occasionally worn by sufferers from allergies or medical conditions. These are sometimes worn round the neck, and occasionally as bracelets.

Finally, all of the above refers to the possibility of you coming across the scene of an accident on a town or country road. If you encounter an accident on the motorway do not attempt to stop unless you are forced to do so through a blocked road. Keep on driving is the normal rule. If you are unsure whether anyone is yet assisting, go on to the next motorway phone box, usually within a mile, park temporarily alongside it on the hard shoulder to report the accident, and then drive on. The greatest danger of motorway accidents is the way that they get bigger through people stopping to watch and get in the way. If the accident is on the other side of the carriageway it is even more important that you do not stop.

Car phones

One practical driving lesson that very few, if any, learner drivers receive before they take their driving test will be on the use of a car telephone, despite the increased use of car telephones in business. The Highway Code lays down some guide lines, but it should be stressed a little bit stronger that the use of 'hands-free' telephones is not just a question of being able to call and talk to other people without removing your hands from the steering wheel.

The real danger associated with the use of car telephones, on the move, is that they are not 'brain-free'. Ideally, telephones should *never* be used whilst on the move by the driver, in any circumstances. If passengers are able to use them to make and receive calls without interference to the driver, that is fine. However, not many drivers fit telephones for the benefit of their passengers. Drivers should never use a car telephone when their attention is necessary on the road ahead. If a call is received, or needs to be made, the driver must pull in and park. If it occurs when on a motorway, the call should be put back until a service area can be reached. Perhaps the only occasion when a call could be made safely whilst still moving is when the driver is held up in a long and often stationary queue of traffic.

In the event of any road traffic incident, if it can be shown that one of the drivers concerned was using a telephone immediately prior to the event, other road users involved would have a very good case to claim driving without due care and attention as a contributory cause. And the police would undoubtedly take action if they saw any driver using a telephone in circumstances likely to create danger to others.

Travelling abroad with your car

Although quite a large number of countries around the world still drive on the left, whenever new drivers think of travelling abroad they naturally assume that they will have to learn to drive on the 'wrong' side of the road. This can be a bit worrying; especially when you realize that it is not the sort of thing that you can practise readily beforehand. Nevertheless foreign travel with your car is not only possible, it is one of the best ways to get to know another country. Travelling by plane or train doesn't quite have the same feel to it, and travelling abroad by coach usually means you are taking your own little bit of Britain with you. In a car you are free to wander as you please, to stop and move on again as the fancy and the scenery take you.

If the thought of travelling abroad really worries you, perhaps you should make your first trip 'overseas' in your car to the Irish Republic. There they do drive on the same side of the road as we do and the rules are very similar. Though comedians might try to convince you that the Irish travel on both sides of the road in order to make all visitors feel at home! There are some small differences between Eire and Great Britain. Learner drivers seem to prefer to put their L-plates in the middle of the windscreens, and there is no requirement for L-drivers to be accompanied. But the volume of traffic and the pace of life is so gentle and genteel that it doesn't seem to matter. About the only rules you need to bother about are take your time, and allow much longer to get from place to place. Not only will this allow you to linger when you find things of interest, it will also prevent you worrying when your timetable seems to have gone by the board.

Getting there by car involves using one of the car ferries, which go from various ports in England, Scotland and Wales either direct to the Republic or via the Ulster port of Larne. Whichever way you travel you won't need a passport; nor will there be any language or money changing problems.

The road signs are very similar, with a few odd, but obvious, exceptions. The rules of the road are almost identical. Speed limits tend to be a bit lower than in Britain; but the only drivers who don't seem to be obeying them are other visitors with British number plates. It is not worth while joining them; far better to take your time and enjoy the scenery and the pleasures of driving in a different world.

If you choose to travel to France or elsewhere on the Continent the trip on the ferry is very similar. You can choose either to take the short trip on the Dover/Calais route; or you might prefer the longer hops from the South coast to Britanny which can even allow you a night's sleep on the ferry before starting your trip across France.

Driving on the Continent presents a few initial fears amongst new, and some experienced, drivers. The most probable reason for this, of course, is the fact that foreigners always seem to drive on the 'wrong' side of the road. In fact that is no problem. Once you are driving along it makes perfect sense to follow the pattern which exists. The only occasions when driving on the wrong side of the road can present any confusion are when you are faced with a large roundabout and you may momentarily want to look or even turn the wrong way, or after you have called into a service station for fuel.

The best way to avoid error is to have one member of the crew whose responsibility is to shout 'keep right' or something similar each time you have a potential problem. Oddly enough, you are less likely to err when driving on your own, as your concentration is much more on what is happening outside; it is only when you are driving a car full of passengers that conversation and discussions can distract you.

France is the most popular of all the continental countries which attract British drivers. This is no doubt primarily because it is the nearest of our neighbours, but driving in France is relatively easy to cope with. The roads in France are generally speaking very straight and not very busy in comparison with their British counterparts. France is a vast country with about the same population as Britain but fewer vehicles. There are about a million miles of roads, including more than 4000 miles of motorways, and about half of the rest are good 'N' roads which are regarded as part of the 'National Network'. Where these N roads do not duplicate the motorway routes they are usually upgraded to very high standards. Certainly they provide visitors with the opportunity to drive fairly long distances relatively easily. The roads are often lined with the poplar trees reputedly planted on the orders of Napoleon to shelter his troops on the march. Overtaking obviously needs care; this is where a sensible passenger is useful. Not to make decisions for the driver, but to tell the driver when a decision can be made.

The motorways in France tend to be boring and they can be expensive to the traveller because of the many péages; but the 'aires' – picnic sites with loos and other facilities – are a very welcome feature. For travellers who do not wish to eat in service stations but simply to rest, wash and use the toilets, they are superb and better than you can find anywhere else in Europe.

Traffic jams are not unique to Britain of course. The 'périphérique' in Paris is notoriously as bad as the M25 and the North Circular combined. French drivers know how to block up even the smallest road when it is time to travel home from work. But, generally speaking, their traffic jams are short lived and are invariably accompanied by continued horn blasting and shouting.

Driving in France is a real delight and will surely tempt all new drivers to return time and time again to savour the delights of pleasurable motoring. Indeed, all the countries of Europe have their

delights; once you have tasted one of them you will realize what
pleasures await you over the water.

Rules for driving in France

- Most road and traffic signs are standardized as in Britain, though
  do note that most French traffic lights are placed overhead (not to
  one side) and contain no amber in the sequence red to green.
  However, the French do drive on the right and overtake on the
  left. Overtaking is not allowed if it entails crossing an unbroken
  line in the road or on the approach to the brow of a hill even when
  it is unsigned.

- Speeding and drink-driving offences are often the subject of on
  the spot fines, and possible impounding of the vehicle where drink
  is involved. Accidents which involve injury must be reported to
  the Police or Gendarmerie. Where damage is caused, but no
  injury, a 'Notice of Motoring Accident' form must be filled in by
  both parties and signed. This then becomes a legal document.

- Seat belts must be worn by all persons where fitted. Children
  under ten are not allowed in the front (except for two-seater cars).

- Hazard warning triangles must be carried, as should a spare set of
  light bulbs.

- It is not compulsory to paint your lights yellow these days, nor to
  fit masks, but it makes sense to try to avoid dazzling oncoming
  traffic.

- Speed limits: every town or village has its name on a board as you
  approach, which marks the beginning of the urban speed limit of
  60 k.p.h (37 m.p.h.) unless a lower speed is indicated. The end of
  the limit is signified by the name of the town with a line through
  it.

- Speed limits on single carriageways are 90 k.p.h. (68 m.p.h.)
  unless the roads are wet. In this case they are reduced to 80 (50).
  Dual carriageways: 110 (68) unless wet 100 (62)
  Motorways: 130 (80) unless wet 110 (68).

- A *minimum* speed limit of 80 k.p.h. (50 m.p.h.) applies on the
  overtaking lanes of motorways in good weather during daylight.

All other variations are signed and the word 'rappel' is used to signify a continuing restriction. A final word on speed limits is that they are rigidly enforced by squads of enthusiastic traffic motor cyclists.

- Priority always used to be given to traffic on the right (the equivalent of the left in Britain) which used to mean that even the slowest old tractor could and would emerge quite blithely on to your road, taking 'right of way'. Nowadays the rule only applies in built-up areas; nevertheless it is as well to remember that not all French tractor drivers know this. The sign giving you priority is a yellow, white and black *diamond*. Look out for it.

Avoid car theft

One of the most expensive lessons that any new driver can learn is failing to lock his car. Over a million motorists every year have either their vehicles stolen or valuables taken from them. Although professional car thieves are not always deterred by locks, burglar alarms and the like, there is no doubt that the majority of car thefts are committed on the spur of the moment. A door handle is tried and if the door opens it is only the work of a moment to take anything of value for the opportunist car criminal. If the keys are left in the ignition, it is even easier to drive the car away where it can be systematically stripped clean.

All of this is easily preventable. Never leave your vehicle unlocked, or with anything of value in sight, even if the vehicle is left out of sight for only a few moments. When you park your car in a car park make sure that you have locked all the doors and the boot, and that your shopping, coats, cameras and similar tempting items are all stowed away.

These precautions will never deter the professional car thief of course. If you are driving a particular type of vehicle – Sierras and Cavaliers are quite popular, so are Porsches and Mercedes – which is in demand by car thieves, then they will arrive equipped to cope with door locks, steering locks, anti-theft devices and noisy alarms. But these thefts are relatively rare and very difficult to prevent.

The most common vehicle thefts are opportunist, and these are the ones which you need to guard most against.

There are many professional anti-theft alarms available these days. Some of them are intended to deter thieves from opening the doors or breaking into windows; others are intended to prevent the car from being driven away. But these only work if they are switched on. Many of the manufacturers of such alarms supply stickers to say that the vehicle is fitted with them.

The basic rules of leaving your car as secure as possible against car thieves are simple and should be put into practice every time you leave your vehicle. They are:

- Never leave the keys in the ignition switch;
- Always check that all doors and the boot are locked;
- Never leave your spare keys in the car, even if you think they are hidden in a safe place;
- Take great care when you choose a parking place. Avoid badly lit or very quiet places if you can. Choose well-lit popular car parks;
- Never leave any items on view, even jumpers, anoraks, shopping bags and parcels will attract the opportunist thief;
- Have your vehicle registration number etched onto all of the windows. Many insurance companies will actually pay you to have this done;
- Fit an obvious anti-theft device. The sort that locks the brake pedal to the steering wheel, or clamps over the parking brake and gear lever, is ideal. If you have an electronic alarm, use one that prevents the vehicle from being driven away rather than one which just makes a noise. Many passers-by just ignore car and burglar alarms these days.

Remember that, statistically, if you live or work in a busy town or city, you stand a 'one-in-four' chance of having your vehicle broken into or stolen. You stand a much better chance of not having this happen if you take precautions.

Car wash

One of the joys of ownership of a motor car is the pleasure of driving a nice shiny, clean, car. Unfortunately this means that it has to be washed and cleaned regularly. Some drivers actually enjoy this chore and take great pleasure in seeing their pride and joy restored each week to pristine showroom condition. Keeping a car clean not only looks good, it actually helps to maintain its value on the secondhand car market, so a weekly wash-and-brush-up can be looked upon as an investment.

However, many drivers prefer to use a commercial car wash. Certainly the costs involved are minimal, and a compromise can often be achieved which satisfies both those who feel too lazy to clean the car themselves and those who like to do it themselves but have insufficient time. This involves putting the car through the car wash to remove the initial dirt and grime and then finishing off the rest of the car by hand at leisure.

Quite often it is the thought of going through a car wash for the first time which puts new drivers off. They have seen many films, or television adverts, of windows staying open, aerials and wipers being torn off and similar indignities being inflicted whilst the drivers remain locked helplessly in the machinery. The worries of these and similar incidents is enough to make them scared to try.

In fact most car wash mechanisms are not only foolproof, they are usually idiot-proof too. What happens is that you go to the kiosk to pay for a 'token'. Quite often there is a choice of three or four different levels of wash. Ideally, for the first time the new driver should go for the cheapest and simplest wash. The car is then driven up to the entrance of the car wash and stopped so that the driver's door is level with the slot which accepts the token. The car is then driven forward to the next stop where a green button needs to be pressed. Then, and only then, the driver should close his own window and switch off the engine. This especially applies if you have electrically controlled windows. It can create a panic if you have electric windows and only remember them *after* you have switched off the engine.

There is usually still plenty of time to check that doors are all closed securely and windows – including the sun roof if one is fitted – are right before the machinery actually reaches the car. Inside the car you can feel that you are moving forward at first but you soon realize this is only the sensation caused by the washing machinery moving round the car. Generally speaking most windscreen wipers and roof aerials can cope with automatic car washes quite well; but if there is any doubt make sure they are as flat as possible before you enter.

Don't operate the screen wipers yourself until after the brushes have passed over, from the front to back, and then from the back to the front again. Only when the machinery comes to rest should you switch on your engine again and move slowly through the rest of the gate until you are clear of the car wash. Then stop and walk round the car to dry off the windows and raise the aerials again if need be.

Accident avoidance is your aim

Eighty-five per cent of all road and traffic accidents have road users as their sole cause. This isn't surprising really, when you bear in mind that accidents are unlikely to happen unless a road user is doing something which creates the problem first of all. Regrettably, cyclists and pedestrians, especially children and the elderly, are amongst those who come off worst in accidents. The reason for this is obvious.

As a driver, especially a new driver, you need to be extra careful with all other road users when you are in charge of your vehicle. As you learn to drive you will realize that there is a whole list of new rules about driver behaviour that you are required to learn. In each case there is a sensible and safe reason why you should do something. In fact every lesson you take whilst learning to drive is designed to improve your skill and make you safer. A skilled safe driver will not only pass the driving test but will also remain skilled and safe for years to come. Accidents, especially road traffic accidents, are badly named, because it makes it sound as if it was no one's fault. Accidents are in fact caused, and should be referred to as 'incidents'.

You as a driver have a responsibility to everyone else on the road not to do anything which could involve them in one of your incidents. So when you are practising a new skill, or learning a new part of your driving programme, remember that you are being taught how to behave safely.

Starting, driving along following a safety line, and pulling in to stop are not terribly difficult skills to learn, which is why they are the first things your instructor will teach you. They are also fairly safe exercises to carry out. It is only when you want to cross other people's paths that you need to take extra care. Watch a stream of traffic moving along the road, and you will see that there are two factors to take into account: the speed at which they are all travelling, and the distance between each vehicle. Provided the speed remains constant and the distance between each vehicle is kept at about two seconds or so, there should never be any traffic incident. It is only when one or more of the drivers tries to travel faster than the rest, or closes the gap from the preceding vehicle, that problems can arise.

Therefore one of the very first safety lessons you will learn is always to drive at the same speed as the rest of the traffic; that way people will be less tempted to try to overtake when it is not safe. Similarly, you are always responsible for the gap between yourself and the vehicle in front. Ideally a 'two-second' gap is what you need; but in some circumstances it may be better to create a slightly longer one. Get your instructor to test your two-second interval quite often in your early lessons, and make sure you always try it out on yourself for the rest of your driving life. Don't wait until traffic conditions get bad before you start to do it. Say it to yourself whenever you are driving along just to get used to the idea.

'Only a fool breaks the two-seconds rule', takes two seconds to say. If you have time to say this, from the time the car in front passes an object until you reach it, you have the right distance. When driving on dual carriageways at high speed, you can always change it to 'Only a fool breaks the five-second rule, with a little left over for the other fools'.

Accidents involving moving motor vehicles come under three classifications:

- nose to tail shunts, which are the most common;
- turning right incidents; and
- overtaking with insufficient clearance.

Fortunately it is the least common ones, overtaking, which are the most dangerous. The reason for this added danger when meeting an oncoming vehicle is that the resultant collision occurs at the combined speed of both vehicles. Turning right is less dramatic but nevertheless the cause of quite a lot of road traffic injuries and accidents. Nose to tail shunts quite often do a lot of vehicle damage, but are not always so painful to the occupants of the cars.

All these accidents are avoidable however.

Nose to tail Avoid being fourth in a queue at any time. If the traffic ahead of you is forming a nice long queue, remember that you need a slightly longer gap between you and the vehicle immediately in front. That way you are not fourth in the queue any more, but first in the new one. Can you see the logic? When something happens up ahead, the first vehicle in the queue reacts and then brakes. The second car is slower in reacting and needs to brake harder. The third car is faced with an emergency stop. The fourth car doesn't have time to stop; it and the following vehicles then pile into each other. Instead of being fourth or fifth, create a bigger gap, and be first in your own queue.

Turning right You cannot be in danger when you turn right if you make sure that you can complete your right turn *before* you start to turn. The reason some drivers and motor cyclists have problems when turning right is that they start the turn before they have checked that it is clear. Make a point of never starting your right turn until you can see clearly into the new road. And never turn right while pedestrians are crossing the road you are entering. Wait until it is clear.

Overtaking This is the easiest of all incidents to avoid. Why bother to overtake in the first place? Only ever overtake if you have a guarantee that it is perfectly safe to get past, and return to your safety line, without inconveniencing anyone. The very definition of overtaking is to get past another moving vehicle. Unless there are good reasons to get past a slower moving vehicle, wait until you are sure that it is necessary, safe and convenient.

Every year in Britain over five thousand deaths, and fifty thousand injuries, occur on the roads needlessly. Most of them could be avoided if only drivers, riders and other road users thought out the result of their actions before committing themselves.

While on the subject of road accidents, it is worth while stressing one factor which all too often is avoided, because it doesn't bear thinking about, and that is the actual forces involved when an impact of any kind takes place. Because you will spend most of your motoring life sitting in a nice, warm, cosy compartment of a motor car, you can feel divorced from the real world. However, at a speed of 15 m.p.h. any sudden stop is the equivalent of being dropped face downwards six feet on to a concrete floor. At a speed of 30 m.p.h. this distance is increased according to the inverse proportion law, so that the impact is now the equivalent of being dropped from the roof of a two-storey house: about 35 feet. At 60 m.p.h. the sudden stop is now the equivalent of being dropped from the fourteenth storey of a block of flats, 172 feet.

These figures should remain at the back of your mind whenever you sit behind the wheel of a motor car. With the added reminder that two cars, both travelling quite lawfully at 30 m.p.h. in a built-up area, will have that impact speed of 60 m.p.h. if they meet head on.

You do not have to be breaking the speed limit to cause very serious injury or death when on the road.

Questions

Part Five

1 You are driving on a motorway and you see flashing amber lights ahead. This means that

(a) You must prepare to stop

(b) You must slow down to 30 m.p.h. or less, unless told differently

(c) You must move to another lane

(d) You must use your flashing hazard indicators to warn others

(e) There may be an unseen hazard further ahead

2 You are intending to make a driving trip lasting more than 200 miles for the very first time. Before setting off you should

(a) Check you have enough fuel for the journey

(b) Make sure you have planned the route

(c) Make sure your vehicle has recently been serviced

(d) Take out special travel insurance

(e) Ensure your passengers are comfortable

3 You are intending to buy a caravan to tow behind your own family car.

(a) You can practise towing it whilst still a learner driver

(b) You need to take a special driving test

(c) You are limited to a maximum speed limit of 60 m.p.h.

(d) You must fix extra mirrors

(e) Reversing is exactly the same as with a normal car.

Answers

1(a) If you see amber flashing lights on a motorway you must always be prepared to stop or slow down. Find out the reason for the flashing lights as soon as you can.

(b) Unless the flashing amber lights are accompanied by some other indication the simple meaning of them is to slow down to 30 m.p.h. or less.

(c) There may be no point in changing lanes, unless you are told to do so. In any case you should be in the left lane unless you are overtaking. Flashing amber lights tell you not to overtake unless you can see it is clear and safe to do so.

(d) You should not normally need to use your own flashing hazard warning lights. Other traffic can also see the amber lights on the motorway. However if there is any specific danger and you wish to inform others, your hazard warning lights may be used to do so.

(e) The purpose of flashing amber lights on motorways is to enable the authorities to give advance warning of unseen hazards ahead, therefore even if you cannot see any dangers drive slowly until you know it is safe.

2(a) It is not always possible to check you have enough fuel to complete a long distance journey. Nevertheless your preparations should include knowing how much fuel you have and ensuring you can obtain and pay for fuel on the journey if needed.

(b) Certainly it is essential to plan the route. Ideally you need a front seat passenger who can remind you which road you need to take. If you are driving on your own, write out the list of main towns and road numbers to keep alongside you whilst you are driving. Plan your route to include a number of comfort stops on the way. Ideally never drive for more than an hour without stopping.

(c) Your vehicle should be regularly serviced. If you don't know when it was last done, or if a service is needed, arrange to have one done before your trip.

(d) There is no need for any special insurance cover, but it is well worth confirming with your insurers exactly what your total insurance cover is, and especially to find out what you are not covered for.

(e) It is very important to ensure that your passengers are as comfortable as possible whenever you drive on any journey. And even more so on long journeys.

3(a) You are not allowed to tow a caravan or any trailer while holding a provisional driving licence. Nevertheless if you hold a full licence, don't be scared to ask for professional lessons and advice before driving a strange vehicle, towing a caravan, or driving in any unusual circumstances.

(b) There are no special driving tests for drivers of cars towing caravans. However the current requirements of the European driving licence places a restriction of 14 cwt on the weight of a caravan. Any trailer weighing more than this would require an articulated licence, and this does demand a special driving test.

(c) There is a maximum speed limit of 60 m.p.h. on all caravans, unless the caravan itself has a double axle with four or more wheels. If you are towing a caravan and you are conscious of a build up of traffic behind you it is both courteous and common sense to pull in to allow them to pass whenever conditions are suitable.

(d) Although there is no legal requirement to fit extra mirrors, it makes sense to do so. Try to ensure you have maximum rearward vision, including the use of periscope type mirrors too. However, the fitting of mirrors is not enough; they also have to be used intelligently.

(e) Reversing a car with a caravan attached is a skilled task. The initial movement requires some study. Ideally get professional help and tuition. You will certainly need to get plenty of practice in before you attempt to reverse into any confined spaces.

Index